The Political Geography of Conflict and Peace

The Political Geography of Conflict and Peace

Edited by

Nurit Kliot and Stanley Waterman

Belhaven Press
(a division of Pinter Publishers)
London

First published in Great Britain in 1991 by
Belhaven Press (a division of Pinter Publishers),
25 Floral Street, London WC2E 9DS

British Library Cataloguing in Publication Data
A CIP catalogue record for this book is available from the
British Library

ISBN 1 85293 133 7

Library of Congress Cataloging in Publication Data
The Political geography of conflict and peace/edited by Nurit Kliot,
 Stanley Waterman.
     p.   cm.
   Includes bibliographical references and index.
   ISBN 1-85293-133-7
   1. Military geography. 2. Geopolitics. 3. World politics–1945–
I. Kliot, Nurit. II. Waterman, Stanley.
   UA990.P67   1990
   327.1′6–dc20
                                                90–48866
                                                    CIP

Typeset by Witwell Ltd, Southport
Printed and bound by Biddles Ltd of Guildford and Kings Lynn

Contents

List of Contributors

Gerald Blake Collingwood College, University of Durham, South Road, Durham, DH1 3LT, UK.

Kenneth E. Boulding Research Associate and Project Director, Institute of Behavioral Science, University of Colorado, Campus Box 484, Boulder, Colorado 80309–0484, USA.

Naomi Carmon Faculty of Architecture and Town Planning, Technion, Haifa 32000, Israel.

Saul B. Cohen Governor's School and Business Alliance, 11 West 42nd St., 21st Floor, New York, New York 10036, USA.

Henning Heske Institute of Geography, University of Dusseldorf, Universitatsstrasse 1, Dusseldorf 4000, Germany.

Baruch Kipnis Department of Geography, University of Haifa, Haifa 31999, Israel.

Nurit Kliot Department of Geography, University of Haifa, Haifa 31999, Israel.

H. van Korstanje Social Geography Institute, University of Amsterdam, Jodenbreestraat 23, Amsterdam 1011 NH, Netherlands.

Anthony Lemon School of Geography, University of Oxford, Mansfield Road, Oxford, England.

Alexander B. Murphy Department of Geography, College of Arts and Sciences, University of Oregon, Eugene, Oregon 97403–1218, USA.

David Newman Department of Geography, Ben-Gurion University of the Negev, PO Box 653, Beer Sheva 84105, Israel.

John O'Loughlin Institute of Behavioral Science, University of Colorado, Campus Box 484, Boulder, Colorado 80309–0484, USA.

Geoffrey Parker University of Birmingham, School of Continuing Studies, P.O.B. 363, Birmingham B15 2TT, UK.

George Quester Department of Government and Politics, University of Maryland, College Park, Maryland 20742, USA.

Dennis Rumley Department of Geography, University of Western Australia, Nedlands, Western Australia 6009, Australia.

Peter J. Taylor Department of Geography, University of Newcastle-Upon-Tyne, Newcastle-Upon-Tyne NE1 7RU, UK.

Stanley Waterman Department of Geography, University of Haifa, Haifa 31999, Israel.

Mark Wise Department of Geographical Sciences, Polytechnic South West, Plymouth, Devon PL4 8AA, UK.

Herman van der Wusten Social Geography Institute, University of Amsterdam, Jodenbreestraat 23, Amsterdam 1011 NH, Netherlands.
Oren Yiftachel Department of Geography, Planning Department, Curtin University, PO Box V1987, Perth, Western Australia, Australia.

Preface

This collection of 15 papers written by 18 contributors is the outcome of an International Conference on *Geography, War and Peace* which was held at the University of Haifa in January 1989. This conference was the eighth in a series of conferences sponsored and supported by the Commission on the World Political Map of the International Geographical Union. This conference was the first to be held after the group's status as a Commission was approved and the second to have been held at the University of Haifa, the first having taken place in 1982.

The conference had both an international and an interdisciplinary character. The 60 participants represented 13 different countries, and varied disciplines such as geography, economics, political science, and behavioural sciences. Some of the sessions attracted up to 150 participants.

The editors would like to express their gratitude to all the authors who contributed their papers to this book; to the University of Haifa for its support in the preparation of the manuscript; and to Belhaven Press for their cooperation in producing this volume.

1 The Political Geography of Conflict and Peace – An Introduction

Nurit Kliot

Introduction

At a time of rapid, perhaps revolutionary, political changes in the international system it is extremely difficult to introduce a collection of papers dedicated to geography, war, and peace. The pace, intensity and inter-relations of the political events, witnessed by each and every individual on their TV screens in their living room, turned almost every person into an active participant in the process of history-making. But is this really the end of history (Fukuyama, 1989)? Can we anticipate that political and geopolitical concepts and theories of the world system will be swept away and replaced by new concepts and theories?

A first assumption is that we lack the historical perspective to judge and evaluate the revolutionary changes taking place in the Soviet Union and in Eastern Europe or the impact of the liberation of Nelson Mandela and the legitimization of the ANC mainly because these are just the very first stage in a long process. Second, we may assume that the political process of revolutions, democratization, economic reforms and adjustment to a new political world have destablizing effects on the international system. Consequently, conflict and warfare will continue and perhaps increase for a while at least.

Let us examine our changing world along two sets of dividing lines, those between East and West, on the one hand, and along the North–South divide, on the other. At the same time we will also focus on change and continuity within the world political system in addition to looking at stabilizers and destabilizers in this system.

World system, power and geopolitics

The first observation is that we live in a period of international system change and a system change can be classified as a stabilizer. The most important system change concerns the East–West axis, namely the eradication of the bipolar structure of the world and the dissolution of superpower rivalry and the emergence of a multipolar structure of 'a world of many consortia in which no single actor can dominate the rest, a world of forging coalitions and agreements' (Williams, P., 1989; Inoguchi, 1989: 22; Kaiser, 1989B: 209).

The hierarchical multipolar structure of the world system comprises two geostrategic regions: the trade-dependent maritime world based on the USA and the Eurasian continental world based on Russia. Each geostrategic region contains several geopolitical regions. Between these two principal geostrategic regions are the zones of rivalry between the two, termed by Cohen (1973) 'shatterbelt zones'. The ascendence of the multipolar world is reflected in the rise of Europe, Japan and China to the status of major powers. But even more important is the rise of second-level regional powers in the 1970s (Cohen, 1982). According to Cohen the international system has a five-fold complex hierarchical structure based on various components of power. The rise of powers other than the USA and the Soviet Union, such as Japan and Western Europe, has a cooling effect on superpower rivalry and facilitates superpower cooperation. It is important to note that the world order of the past 40 years was reshaped radically after World War II whereas the present system changes are mostly gradual and slow (Masefield, 1988). What is changing slowly is the bihegemonic world order. Hegemony is a condition in which one state is powerful enough to maintain the essential rules governing interstate relations and is willing to do so (Russett, 1985: 213). There have been three such hegemonic cycles. The first hegemonic power was the Netherlands (seventeenth century); this was followed by Britain in the eighteenth and nineteenth centuries and the USA (mid twentieth) (Taylor, this volume, Chapter 6; Taylor, 1989). A large volume of political analysis deals with either 'Pax Americana' on the one hand, or with the USA's decline as hegemonic power on the other. The decline of the USA in the 1970s and the 1980s is related to its defeat in Vietnam and the economic indebtedness that accompanied this era. A huge budget deficit resulting from enormous defence expenditures led to a significant drop in the value of the dollar and to additional difficulties in trade (Masefield 1988).

The grand strategies for the American policies which evolved after World War II were founded on the self-perception of the US as a global superpower. This role called for American intervention and involvement throughout the globe in order to maintain the balance of power and to contain Soviet influence and power. This strand of writing is still frequent. 'Pax Americana', in which America retains its leading position in the world, is still the most favoured world order in Japan because it enables Japan to keep its traditional economic role, with no drastic increase in its security role which is largely delegated to the US (Inoguchi, 1989: 20). But most of the present grand strategies are based on 'asymmetric' foreign policy in which some regions such as Western Europe, the Persian Gulf and Northeast Asia are defined as areas of 'great intrinsic importance' (George 1988: 270; Walt, 1989).

America's role in the Third World is one of the issues on which 'grand strategists' differ. The balance sheet shows that in development aid to Third World countries, America's role has been characterized by a marked decline (Kaiser, 1989b). Yet, according to one view, American interests in the Third World are too important and too numerous to ignore, whereas there is little or no justification for American commitment in the Third World, according to another view (David, 1989; Walt, 1989: 33). O'Loughlin (this volume, Chapter 3) provides us with the historical development of grand strategies based on the 'symmetric' or 'asymmetric' world views which guided American policy-makers after World War II.

Within the framework of hegemonic order, the American decline is described in a broad body of literature mostly written at the end of the 1970s (Rosecrance, 1976; Oye, 1979). Most of the responses were that it is a gross overstatement to consider the US as 'ex-hegemonic' (Strange, 1987; Russett, 1985; Kaiser, 1989b: 209).

Grand strategies based on superpower rivalry and containment led the world to a 'peace' maintained by deterrence and cold war (Russett, 1985, 217; Taylor, this volume, Chapter 6). But there is also evidence that periods of hegemony produce a higher incidence of wars, many of which have taken place in Third World countries (Mansfield, 1988, p. 37; Russett, 1985: 216). In either a bipolar or multipolar world, a decline is also noticed in the hegemonic power of the USSR, mainly as a result of both internal and external problems. Internally, the weak performance of the Soviet economy was the main reason for Gorbachev's economic and political reforms (Kaiser, 1989a; Newsom, 1989; Kaiser, 1989b). Also, since 1986, manifestations of nationalistic unrest have occurred in the Soviet Union on a scale unprecedented in Soviet history and at a pace not previously known (Smith, 1989: 8; Mathisen, 1984).

Toqueville's law has taught us that the most dangerous moment for an oppressive regime is when it starts to reform and particularly when, in the process, rising expectations are frustrated and disappointed. The success or failure of the reforms in the Soviet Union will have enormous impact on the international system. As it appears from the perspective of March 1990, the present situation in the USSR is a destabilizing one. There is a danger that the opposition emanating from Party apparatchiks and bureaucrats may succeed in ousting Gorbachev while there is also a danger that the USSR might be fragmented by the rise of nationalist movements. It is likely that this process will be accompanied by conflict and warfare.

In the multipolar world the other major actors are Japan, China, (Unified) Germany and the rest of the European Community. Today, Japan is one of the world's economically most powerful states; so far it has shown little inclination to become a world military power again. Japan is the focus of a regional integration of the 'Yen Zone' and has taken upon itself to serve as a link between the US economy and the Asian Pacific economies (Kaiser 1989a: 21; Inoguchi, 1989: 20–4; Masefield 1988: 7; Cohen 1982). But Japan has become more involved in the world system: in 1988 Japan took over from the USA as the leading donor of development aid to Third World countries (Kaiser, 1989b: 213). China, a candidate first-order power, showed in its brutal suppression of Tian An Men protesters that continuity of the authoritarian regime can be anticipated. The crushing of the movement for democratization created a deterioration in the close relations that had existed between China and the West for some years. The Tian An Men massacre had a clear destabilizing effect on the world system.

More than Japan, a reunified Germany is probably the only nation which can match the superpowers in terms of combined economic, political and military might. There are two main scenarios for the role to be played by a unified Germany. The first places it at the head of a continental grouping of central European states; the second incorporates Germany into a New Europe in such a way that it no longer seems to be a threat to its neighbours (Mathisen, 1984; Masefield 1988; Kaiser 1989a: 135). Perhaps, neutrality for reunified Germany and perhaps demilitarization for all or part of Germany will be two of some of the solutions that will be suggested by some European nations which might like to curtail the power of re-unified Germany.

Finally, the European community is taking on a more internationally political nature with the development of closer coordination of foreign policy (Wise, this volume, Chapter 8). Another development which could be expected is that the traditional alliance between Europe and the USA will be weakened as a result of the detente between East and West and the fewer common interests in the security of Europe. The Europeans have begun to come to terms with the idea that closer cooperation among states means the loss of some of each state's exclusive sovereignty. Europe in 1990 appears to be more amenable to neutralism and non-alignment, and is characterized by the horizontal integration of the economic and political relations of the EC more than by vertical integration (George, 1988; Kaiser, 1989a,b; Newsom, 1989).

The growth of the European Community raises questions about the future of the USA – European relationship and the future of the NATO Alliance, especially in the light of events in Eastern Europe (Newsom, 1989). Eastern Europe is extremely attracted to the economic and technological benefits which might be gained from joining the EC but Wise (this volume, Chapter 8) does not believe that the EC can be quickly expanded to absorb the Eastern European countries. Wallerstein (1983: 142) predicted a major geopolitical realignment over the next 20 years which will decrease the probability of global war. He predicts two axes: a Beijing–Tokyo–Washington axis versus a Moscow–Bonn–Paris alignment. At present, it seems that there is more promise in the European axis developing than in the Pacific one.

War, detente, deterrence and disarmament

Superpower empire-building, superpower rivalry and the policy of contain-
ment extended the frequency of conflicts and the involvement not only of
the US and the USSR in warfare, but also of Third World states (van der
Wusten, 1985; Mansfield, 1988: 14; O'Sullivan, 1985: 32; Douglas, 1985).
Perhaps the most important stabilizing change in the relationship between
East and West is the widespread faith in detente and disarmament.

Deterrence has been achieved by the build-up of a huge arsenal of nuclear
missiles. The proposition that nuclear weapons have prevented the outbreak
of war between states possessing nuclear weapons is confirmed by the fact
that such countries have actually been entangled in conflicts with countries
without nuclear weapons (the US in Vietnam; the USSR in Afghanistan).
Nuclear weapons have become an instrument of politics and this instrument
will continue to apply in a Europe where conventional disarmament has
achieved its first successes (Kaiser, 1989a; Newsom, 1989). According to this
view, the more drastic the reduction in power the more the West will
continue to depend upon a nuclear minimum as a safety net for preventing
conventional war. This is accompanied by a discussion of establishment of
nuclear weapons free zones and the withdrawal of tactical nuclear weapons
from Central Europe. This has become an important foundation of German
policy and probably will be reinforced by the reunification of Germany and
the alignment of a neutralized and demilitarized Middle Europe. In the field
of East–West disarmament, amazing developments have occurred. After 16
years of unsuccessful mutual balanced force reductions (MBFR) a new
phase has begun.

The East has accepted the principle of asymmetrical reductions where
imbalances exist, and concurs with the West in the goal of eliminating the
capacity for large-scale offensives and surprise attacks. In the INF Treaty
and in the Stockholm Agreement, NATO and the Warsaw Pact accepted an
unprecedented degree of mutual on-site verification, including supervision
of the production and inspection of their arsenal reduction. Today, many
believe that nuclear arms in Western Europe can be reduced and that the
chances for the second phase of detente have never been better.

One worldism?

' "One worldism" is the coming together of three ideas. First, that we are
economically one world; second that promoting international political
cooperation is making us one world; and third that the fragility of the
environment is making us one world' (Taylor, 1989: 211). 'One worldism'
stands in contrast to a bi-polar world system and assumes a global unity. In
effect, it is also a myth mainly because of the North–South divide. The
North–South gap has widened in the past two decades and, together with the

national and ethnic revival, constitute the most important destabilizer or threat to the world system.

The clash of interests between the poorer developing states located mainly in the southern half of the globe and the richer industrialized states in the north has assumed considerable importance in world politics, and concerns a more equitable distribution of wealth in the world (Mathisen, 1984; Douglas, 1985). The North–South divide has been framed in terms of a world economic core (the industrialized north) and economic periphery (the poor south) (Short, 1982).

The book, *World in Crisis* (Johnston and Taylor, 1986) portrays very well some of the major crises of the developing Third World nations such as food shortages and hunger, scarcity of cheap energy sources, the destructive depletion of natural resources and overpopulation. Together with these so-called crises, the Third World is plagued by huge waves of refugees either resulting from war or drought that their host countries find impossible to support (*The Economist*, 23.12.1989).

The image of the Third World in the public opinion of rich countries is a confused combination of disorder and abuse, absence of democracy, tremendous population growth, illegal migrant labour, war and Marxist revolutionary regimes (Bertrand, 1988: 249). Unfortunately war-making and state-making often coincide among Third World nations. Small and Singer (1982) listed 58 'Third World' wars during the period 1945–80. Since 1987 many Third World wars such as those in Afghanistan, and Western Sahara or that between Iran and Iraq have ended. Elsewhere, such as in Cyprus, the two opposing sides are negotiating over its future (Urquhart, 1989).

But we are reminded that stable peace has not been achieved to anything like the same degree in the Third World as in the First World. Virtually all the wars since the end of World War II have been fought on the territories of Third World states (Russett, 1985: 216). Client states in the Third World have proved enormously difficult to control, especially in the Middle East.

Few political analysts call for the preservation of American interests in the Third World, but many more call for the abandonment of American involvement in the Third World (David, 1989; Desch, 1989). At the same time, as Soviet involvement in the Third World appears to have become unbearably costly, we can also anticipate a retreat from Soviet involvement in Third World countries (Kaiser, 1989b).

The second tenet of 'One worldism' is the unity imbedded in the environment. Environmental issues have become a highly prominent global focus. Acid rain, the depletion of tropical forests, and the protection of the ozone layer have become global problems forcing global cooperation which perhaps, like other forms of international cooperation, challenges the concepts of national sovereignty. But the efforts to resolve environmental problems also risk exacerbating the split between the northern industrialised nations and the poorer nations of the south. Many poorer countries claim that the industrialized world is most to blame for the 'greenhouse effect' and that it now wants the 'have nots' to pay the price for stopping it. Moreover, limits on energy use would cripple many economies in the Third World.

Progress in global transport and the growth in world trade are also considered as supports for global integration. American, European and Japanese multinational corporations which are engaged all over the world reinforce this interaction and integration (Mathisen, 1984). Another area in which 'One worldism' is perhaps emerging and even challenging the sovereignty of some states is the area of data flows created by television, computers and other means of mass communication (Newsom, 1989).

But the most significant development of 'One worldism' is the evolving world economic and financial systems. Side by side with the vertical integration of economic regions such as the Dollar Zone, the European Zone and the Yen Zone, there is also horizontal integration of world markets and economies which cuts across the integration on a vertical or regional basis (Masefield, 1988). The new links between multinational corporations, banks and countries have been strengthened (Thrift, 1986). 'Black Monday' showed that there was, in effect, one global financial system. International agreements – such as GATT, the role of the IMF in negotiating and rescheduling the international debt of Third World countries, and food agreements – indicate the emergence of global solutions to global economic problems (Corea, 1986; Patterson, 1986; Samuels, 1986). The final component of 'One worldism' is the political organization of the world system. The idea of a world system based on political consensus among nations is an old one and it is manifested in the establishment of the League of Nations and the United Nations. The role of the United Nations in keeping the peace in Lebanon, Israel, and Cyprus, and its negotiating role in solving other conflicts has increased significantly in the last decade (Urquhart, 1989). There are other internationally negotiated orders such as the Law of the Sea Treaty, the International Telecommunication Union or the Non-Proliferation Treaty, GATT and the IMF (Henrikson, 1986).

However, vertical integration, as represented by regional supranational organizations and by hegemonic regional organizations is still the most important form of the international order. First, there are the imposed hegemonic organizations such as the Warsaw Pact and negotiated hegemonic organizations such as NATO.

We find that supranational regional organizations in every continent – OAS, ASEAN, and the European Community (the most successful) – reflect this trend. Regional subsystems are characterized by their functions and by the geographical proximity of the countries involved. The countries are also similar in their political and economic needs (Thompson, 1973; Vayrynen, 1984). Based on membership in IGO, Nierop (1989) identified eight regional clusters of which the most prominent were Eastern and Western Europe, British Caribbean, and British Africa. This structure of the world system reinforces the multipolarity of the international network and thus contributes to a more stable world. On the other hand the structure of the diplomatic network still reflects primarily the political power orientations of the postwar world based on the old hegemonic bipolarity (van der Wusten, this volume Chapter 7).

Images, perception and the international system

Perception is crucial in the process of recognizing and evaluating military power, especially within the framework of the balance of power between the USA and USSR. Luttwak was able to show that the overall military power of the USA manifest to all allies and clients, is the sum of separate perceptions among American policymakers of different 'power zones'. Differentiation in perception is typical also to allies and clients who do the 'perceiving' of the American role in the world (Luttwak, 1984: 16–17). Perception and misperceptions form important elements in the decision-making processes of small European nations such as The Netherlands. Dutch perception of world superpower rivalry fluctuates between the image of a 'Russian threat' and a need for an American security guarantee to Western Europe against this threat, on the one hand, whereas a more relaxed image of Russia in the light of its efforts to achieve detente evokes associations of moving forward even when practically no progress has been made, as was the case during the Belgrade Conference of 1977–8.

Students of detente expressed their fear that this concept was risking its becoming merely the reflection of an illusion, namely, negotiating nations do not necessarily experience a deterioration in their mutual relations (Alting von Geusau, 1984: 110). The recent agreements between the USA and USSR on the mutual reduction of their armies and nuclear and chemical weapons could point to the positive role played by the detente and the image of 'thaw' between the two superpowers on the real decision-making. We find that behaviour depends on the image. The image is built up as a result of all of the past experience of the possessor of the image. Part of the image is the history of the image itself (Boulding, 1975: 6–8). More than anything else images reflect subjective knowledge. Human decisions are not formed in total knowledge of reality but rely upon perceptions of reality which are always incomplete and inaccurate.

The total image in the political process includes not only detailed images of role expectations such as those concerning the American presidency but also symbolic images such as 'Uncle Sam' or the 'Russian Bear'. The latter are particularly important in the summation and presentation of value images (Boulding, 1975: 110–12). Boulding has identified the most important images as being those which the political community has of itself and those which it has of other political bodies in the international system. People hold national images which are historical images formed mainly by the individual in childhood. It is normally a folk image based on a selective view of the nation. Members of the elite have themselves inherited this image and their political decision-making is often founded on these images; images – which cultivate detailed information not necessarily in an objective manner – govern political behaviour.

The role of image formation as it impinges on the political process has been presented by Brecher in his analysis of Israeli foreign policy-making. Brecher (1972) differentiates between the operational environment which includes political conditions outside and inside Israel and the psychological environment which modifies the information according to the predispositions of the decision-making elite. Extremely important is the concept of the 'enemy' as held by policy-makers such as that reflected in John Foster Dulles' attitude toward the Soviet Union (Holsti, 1972). When decision-makers for both parties to a conflict adhere to rigid images of each other, there is little likelihood that even genuine attempts to resolve the issues will have the desired effect. In this respect, Friedlander has pointed to the important role played by changing mutual images of Begin and Sadat in the peace initiative of President Sadat (Friedlander, 1983).

Images are created, as are many other aspects of human knowledge, through socialization. Territorial socialization is the process by which individuals acquire identification with their political areas (Duchacek, 1970). Governments, schools and mass media shape national images (Muir and Paddison, 1981: 41). Such images of nations towards other nations emerge as an outcome of a complex socialization (Boulding, this volume, Chapter 10). People's images of other nations are frequently oversimplified, stereotyped and distorted. Thus, the British described the Russians as hardworking, domineering, cruel, backward and brave, while they perceived themselves in a very positive light (Buchanan and Cantril, 1953). Luostarinen (1989: 123–38) has shown how 'Finnish Russophobia' has changed over the years. The Finnish image of the Soviet Union as an enemy was gradually abandoned when the war between Finland and the Soviet Union was lost. Many of the organizations which fostered Russophobia were disbanded. Increased official contact, tourism, economic cooperation and the removal of the most anti-Russian elements from school books were the most important factors in fostering a new image of the Soviet Union (Luostarinen, 1989: 131). Today, the majority of the Finnish people are not afraid of the Soviet Union and do not consider it a threat to their security.

Hostile relationships at the international level (and the images and perceptions tied to them) are not everlasting. Over the past two centuries, the USA has fought wars against Britain, France, Mexico, Spain, Germany, Italy and Japan, all of which are currently allies to some degree. But it is also evident that hostile relationships do tend to endure and the image of the enemy is self-perpetuating (Holsti, 1972). Even today, for instance, anti-Israeli propaganda in the Egyptian mass media is working to strengthen the enemy image of Israel in Egypt (Yadlin, 1988).

Enemy images play prominent roles within the conflict between Israelis and Palestinians (Newman, this volume, Chapter 14). An anti-Palestinian posture is also sustained by the Israeli mass media and sometimes even by school textbooks. Negative images of the Palestinians are reinforced by the use of graphics on maps and by conscious use of placenames concerning the West Bank (Newman, this volume, Chapter 14).

In the light of present events in Eastern Europe it is interesting to note what is described as the persistence of traditional images and perceptions which continue to have an impact especially on bilateral and multilateral relations within the region (Korbonski, 1989: 120–1). Old friendships between Poland and Hungary will, most likely sustain the special relationship between those two countries. But we can also anticipate that the traditional antipathy and distrust between Czechs and Poles will have a strong effect on the future relationship between these states. The old enmity between Hungary and Romania has already re-emerged in recent incidents between these two countries. Images and perception have crucial roles, not only in power politics but also in the revival of ethno-nationalism.

Ethnic-nationalism and the world's instability

In addition to the East–West conflict (seemingly decreasing nowadays) and the North–South divide (clearly widening nowadays), the rise of nationalism and ethnic conflict accompanied by warfare is the most important destabilizer of the present world system. The most unstable regions are Eastern Europe and the Middle East. What makes ethnonationalism so dangerous to world stability today is its coincidence with East–West rivalry and with the North–South divide. Moreover, the declining influence of ideology has definitely contributed to a detente between East and West, but the influence of nationalistic ideologies exacerbates and inflames ethno-national conflicts.

Nationalism is an ambiguous concept. In the nineteenth century it was the mobilizing factor for the independence movement in the Balkans, Germany and Italy. During the interwar period, nationalism became a very important ingredient of Nazi and Fascist ideologies (Seiler, 1989). Finally, in the process of Third World decolonization nationalism became the most prominent notion.

Duverger (1964) offered a distinction between peripheral nationalism which is progressive and linked to the struggle for self-government, democracy and socialism, and centralist nationalism – conservative and referring to colonialism, imperialism and anti-democratic strains. There is a tendency to differentiate between traditional nationalism in the developed realm and nationalism in the underdeveloped countries. By contrast, the explosion of national and regional problems in developed countries is still awaiting a theoretical explanation that can be agreed upon. Ethno-nationalism or mini-nationalism is a sentiment of an ethnic minority in a state or living across state boundaries that stimulates the group to unify and identify itself as having the capacity for self-government (Shiels, 1984; Snyder, 1982). Confrontation is over power and resource distribution, cultural and political autonomy, and self-determination.

The main theories on ethnic and nationalist revival concentrate on the relative deprivation perceived by an ethnic group, a collective identity perceived as threatened by the culture of another dominant group, and by internal colonialism and centralism exerted by the capitalist state (Duchacek, 1979: 62; Smith, 1982; Williams, 1982: 120; Thorburn, 1989; Hueglin, 1989).

Regional and ethnic dissent appear as a reaction against the centralizing force of the government. Late capitalism is characterized by economic, social, cultural and political centralization and consequently new forms of territorial marginality emerge (Thorburn, 1989; Tura Solé, 1989). Anti-centralism is the most important subjective common denominator of regionalist aspirations. Regionalism is the spatial expression of the peripheralization of marginality (Hueglin, 1989: 214–19). Many ethnic and nationalist demands have been accommodated by central governments by providing various levels of autonomy. Examples of this are the case with the Basques and Catalans, in Spain, the South Tyroleans in Italy, and Alsatians in France (Krejci and Velimsky, 1981; Ramon, 1985; Kliot, 1989).

Application of control mechanisms on the Israeli Arab population and on Malaysia's Chinese minority is another measure to 'accommodate' minority needs – in this case it suppresses minority demands (Rumley, Carmon and Yiftachel, this volume, Chapter 13; Kipnis, this volume, Chapter 15). The persistence of the policy of apartheid in South Africa, in which black majority demands and rights for self-determination, equality and equity are totally ignored by the white ruling minority, is transfused into this country's international political relationships. Through a policy of destabilizing the economies and security of its neighbours, South Africa succeeds in subordinating countries such as Zambia, Malawi, Zimbabwe, Angola and Mozambique (Lemon, this volume, Chapter 12). The authoritarian communist regimes in Eastern Europe and the Soviet Union were successful in suppressing any ethnic demands or call for self-determination. Today with the weakening of the bonds holding the Soviet empire together, ethno-nationalism is appearing once more as a very powerful force. The nations of Eastern Europe are currently redefining their aims and aspirations and questions can be posed about the territorial integrity of the Eastern bloc in its present form. In Bulgaria, Romania and Czechoslovakia ethnic minorities amount to roughly 10 per cent of the total population. Poland still includes approximately 1 million ethnic Germans. Yet the most immediate problem concerns Hungarians. Approximately 2 million Magyar speakers live in Romania, some 600,000 in Czechoslovakia and 500,000 in Yugoslavia (Korbonski, 1989; Eyal, 1989). The tension between Hungary and Romania in recent years has been very high as a result of Romanian policy of resettling its Hungarian population (Korbonski, 1989). There are some

2 million Romanians in the Soviet republic of Moldavia; there are also approximately 1 million Poles within the Soviet Union. West Germany has recently demanded and obtained a variety of special concessions for ethnic Germans in Poland. Persecutions against the Turkish minority in Bulgaria resulted in the flight of some 200,000 ethnic Turks to Turkey in July 1989. The process of democratization and economic reforms in the Eastern European countries has just begun, and their situation is as yet extremely unstable – especially in Romania and Bulgaria, and it is not clear yet if further fragmentation can be anticipated as a result of secessionist trends. In this respect, 'eastern and central Europe appears to have taken a 70-year backward step into history' (see Polonsky, 1975).

A most volatile situation exists in the Soviet Union. First, the 'fire ring of Islam' is spreading from Pakistan–Kashmir–Iran into the Muslim republics of the USSR – particularly to Azerbaijan, Tadzhikistan and Kazakhstan, and perhaps moving from there to the Albanian Muslims of Kosovo, in Yugoslavia.

Manifestations of nationalist unrest have also occurred in the Armenian enclave of Nagorno-Karabakh in Azerbaijan and in the Baltic republics which remember their past independence vividly (Smith, 1989). The Soviet Federation is a federation in name only, and the slow response of the central government to the republics' demands for more equity will probably result in a new loose form of confederation or perhaps even the secession of some of the republics. Another federation which has failed is the Yugoslav federation dominated by Serbs. Separatist trends prevail among the better-off Slovenes and Croatians (formerly oriented towards Austria-Hungary) and among the Albanians – the poorest of all ethnic groups in Yugoslavia (Korbonski, 1989).

As noted earlier, the Middle East conflict is a major source of instability and, with the 'thaw' in East–West relationships, the continuation of this conflict jeopardizes the world's stability.

The Zionist movement can be classified as a classical irredentist movement of revolutionary ethno-nationalism between 1895 and 1948 when Israel gained its independence. From the late 1950s until the early 1970s the Palestinian ethno-national movement can be regarded as an irredentist movement whose main stated goal has been the creation of the secular non-sectarian democratic state of Palestine (Heraclides, 1989). Probably the dominant view at present is to establish a Palestinian state in the Occupied Territories of the West Bank and the Gaza Strip and the PLO has shifted from the stage of being a classical irredentist movement to being an anti-occupation revolutionary movement, fighting against Israeli occupation. Another facet of this conflict is the standing of those Palestinians who are Arab citizens of Israel who have become highly mobilized by the Palestinian uprising in the Occupied Territories. Most of the Israeli Arabs demand equity and the removal of any form of discrimination against them (Smooha, 1978), a difficult task in a state in which the raison d'être is rooted in Jewish nationalism (Waterman, 1990). There is, however, an extreme Muslim sector within this population which will be satisfied only with the establishment of a Muslim State in place of Israel; some Palestinians both within Israel and outside it will be satisfied only when the 1947 UN Partition Resolution is implemented (Cattan, 1988). While a solution involving two states appears to be in the ascendancy as the most viable solution, no negotiations between the adversaries have started yet (Plascov, 1981; Massoud, 1987). The major contributors to the ongoing Middle East conflict are as follows:

1 Incompatible goals – both sides demand total control of the whole of Palestine, that area ruled by the British Mandate.
2 Each side supports its claim for the land on a total negation of the other party. For, as we have already indicated, negative images play an extremely important role in the Jewish–Arab conflict.
3 The rise of the Right to power in Israel and the paralysis of the present government of 'National Unity' prohibits any step towards solution.
4 Finally, the influence of the superpowers on the conflict could become a positive factor for the first time, as the positions of the USA and the Soviet Union become closer.

The ethno-national revival in Eastern Europe and the probability of a Palestinian State raises again the question of further fragmentation of the world political map and changes in the present boundaries (Cohen, this volume, Chapter 2). Further political fragmentation is expected in Africa and Asia (Mathisen, 1984: 13). Yet, we are reminded that the nation-state is unquestionably a powerful myth (Murphy, this volume, Chapter 9). One of the legacies of World War II has been the disapprobation of any extrastate claim to territory based on purely ethnic grounds. Pushing an ethnic national claim to its logical conclusion could call into question the territorial integrity of most states. But in reality, a set of boundary agreements, both continental and maritime, have been agreed upon by some Middle Eastern countries and illustrate how it has been possible to arrive at a peaceful solution to various problems resulting from the colonial tradition of boundary drawing (Blake, this volume, Chapter 11).

In conclusion, we may point, again, at the most prominent destabilizers in the current world system: it is the present wave of ethno-nationalism, the process of reforms and democratization in Eastern Europe, and the gap between North and South. Some will also say that a Unified Germany is going to destabilize the European system. Among the stabilizers we find the East–West rapprochement and disarmament, the growing multipolarity of the world's system and the expanding horizontal integration through the world's organizations and international agreements. Yet, warfare still plagues the Third World, and this realm together with the Middle East, is going to be high on the agenda of world geopolitics.

The contributors to this volume have once more reflected on the mysterious riddle of war and peace, re-examining the contribution of the geopolitical framework, in particular, forms of 'no war no peace' to world stability. They have looked at the organization of the world, at the complexity of alliances, blocs and treaties in their impact on world peace. In their search for sources of instability and warfare they have returned to traditional bases of conflict such as boundary disputes and ethnonational conflicts. The profile of the world system as portrayed in this volume is mostly of a world far away from stability and peace, a world which still poses serious questions for political geographers to answer.

References

Alting von Geusau, F. A. M. (1984). 'East-West Detente: Perceptions and Policies', in N. Oren (ed.), *Images and Reality in International Politics*. New York: St Martin's Press, pp. 105–11.

Boulding, K. E. (1975). *The Image*. Ann Arbor, Michigan: The University of Michigan.

Brecher, M. (1972). *The Foreign Policy System of Israel: Setting, Image and Process*. New Haven: Yale University Press.

Buchanan, W. and Cantril, H. *et al.* (1953). *How Nations See Each Other: A Study in Public Opinion*. Urbana: University of Illinois Press.

Cattan, H. (1988). *The Palestine Question*. Beckenham, Kent: Croom Helm.

Cohen, S. B. (1973). *Geography and Politics in a World Divided* (2nd edition). New York: Oxford University Press.

Cohen, S. B. (1982). 'A New Map of Global Geopolitical Equilibrium: A Developmental Approach', *Political Geography Quarterly*, **1**, 233–42.

Corea, G. (1986). 'Creating a Framework to Strengthen and Stabilize International Commodity Markets', in Alan K. Henrikson (ed.), *Negotiating World Order*. Wilmington, Delaware: Scholarly Resources, pp. 167–80.

David, S. R. (1989). 'Why the Third World Matters', *International Security*, **14**, 50–85.

Desch, M. C. (1989). 'The Keys that Lock Up the World', *International Security*, **14**, 86–121.

Douglas, J. N. H. (1985). 'Conflict between States', in M. Pacione (ed.), *Progress in Political Geography*. London: Croom Helm, pp. 77–110.

Duchacek, I. D. (1970). *Comparative Federalism: The Territorial Dimensions of Politics*. New York: Holt Rinehart and Winston, p. 3.

Duchacek, I. D. (1979). 'Federalist Responses to Ethnic Demands', in Daniel J. Elazar (ed.), *Federalism and Political Integration*. Ramat Gan: Turtledove Publishing Co., pp. 59–72.

Duverger, M. (1964). *Introduction to the Social Sciences*. London: G. Allen.

Duverger, M. (1972). *The Study of Politics*. New York: Crowell. The *Economist* 23.12.1989.

Eyal, J. (1989). 'Eastern Europe: What About the Minorities?', *World Today*, **45**, (12), 205–8.

Friedlander, M. A. (1983). *Sadat and Begin – Domestic Politics of Peace Making.* Boulder, Colorado: Westview.

Fukuyama, F. (1989). 'The End of History?', *The National Interest*, Summer, 1989.

George, L. N. (1988). 'La Decadencia del Dragon: U.S. Hegemonic Decline and the Future of Inter-American Relations', *International Relations,* **9**, 257–80.

Henrikson, A. K. (1986). 'The Global Foundations for a Diplomacy of Consensus', in Alan K. Henrikson (ed.), *Negotiating World Order.* Wilmington: Delaware: Scholarly Resources, pp. 217–44.

Heraclides, A. (1989). 'Conflict Resolution, Ethnonationalism and the Middle East Impasse', *Journal of Peace Research,* **26**, 197–212.

Holsti, O. R. (1972). 'Enemies Are Those Whom We Define as Such, A Case Study', in Ivo M. Duchacek (ed.), *Discord and Harmony.* Hinsdale, III.: The Dryen Press, pp. 184–94.

Horowitz, D. L. (1985). *Ethnic Groups in Conflict.* Berkeley and Los Angeles: University of California Press.

Hueglin, T. O. (1989). 'Better Small and Beautiful than Big and Ugly', *International Political Science Review,* **10**, 209–21.

Inoguchi, T. (1989). 'Four Japanese Scenarios for the Future', *International Affairs,* **65**, 15–28.

Johnston, R. J. and Taylor, P. J. (eds). (1986). *A World in Crisis?* Oxford: Basil Blackwell.

Kaiser, K. (1989a). 'Why Nuclear Weapons in Times of Disarmament?', *World Today,* **45** (12), 134–9.

Kaiser, K. (1989b). 'A View from Europe: The U.S. Role in the Next Decade', *International Affairs,* **65**, 209–24.

Kliot, N. (1989). 'Accommodation and Adjustment to Ethnic Demands – Mediterranean Framework', *Journal of Ethnic Studies,* **17** (2), 45–70.

Korbonski, A. (1984). 'Image and Reality in International Relations: The Case of Eastern Europe', in N. Oren (ed.), *Images and Reality in International Politics.* New York: St Martin's Press, pp. 112–24.

Korbonski, A. (1989). 'Nationalism and Pluralism and the Process of Political Development in Eastern Europe', *International Political Science Review,* **10**, 251–62.

Krejci, J. and Velimsky, V. (1981). *Ethnical Political Nations in Europe.* Beckenham, Kent: Croom Helm.

Leurdijk, J. H. (1984). 'Images and Reality in International Politics: Security Perceptions in the Foreign Policy of the Netherlands', in N. Oren (ed.), *Images and Reality in International Politics.* New York: St Martin's Press, pp. 70–80.

Luostarinen, H. (1989). 'Finnish Russophobia: The Story of an Enemy Image', *Journal of Peace Research,* **26**, 123–38.

Luttwak, E. N. (1984). 'Perceptions of Power in Retrospect', in N. Oren (ed.), *Images and Reality in International Politics.* New York: St Martin's Press, pp. 9–19.

Mansfield, E. D. (1988). 'The Distribution of Wars Over Time', *World Politics,* **41**, 21–51.

Masefield, T. (1988/1989). 'Co-Prosperity and Co-Security, Managing the Developed World', *International Affairs,* **65**, 1–14.

Massoud, H. (1987). 'Moving Towards Mideast Peace', *International Perspectives*, May–June, 17–20.

Mathisen, T. (1984). *Sharing Destiny*. Oslo, Norway: Universitets forlaaet.

Maurice, B. (1988). 'A New North/South Dialogue', *International Relations*, Vol. IX, May, **3**, 244–256.

Muir, R. and Paddison, R. (1981). *Politics, Geography and Behaviour*. London: Methuen.

Newsom, D. D. (1989). 'The New Diplomatic Agenda: A Government Ready', *International Affairs*, **65**, 29–41.

Nierop, T. (1989). 'Macro Regions and the Global Institutional Network 1950–1980', *Political Geography Quarterly*, **8**, 43–66.

O'Sullivan, P. (1985). 'The Geopolitics of Deterrence', in D. Pepper and A. Jenkins (eds), *The Geography of Peace and War*. Oxford: Basil Blackwell, pp. 29–41.

Oye, K. A. (1979). 'The Domain of Choice', in K. Oye, Donald Rothchild and Robert J. Lieber (eds), *Eagle Entangled: U.S. Foreign Policy in a Complex World*. New York: Longman.

Patterson, G. (1986). 'The Gatt and the Negotiation of International Trade Rules', in A. Henrikson (ed.), *Negotiating World Order*. Wilmington, Delaware: Scholarly Resources, pp. 181–98.

Plascov, A. (1981). 'A Palestinian State? Examining the Alternative', *Adelphi Papers*, No. 163. London: International Institute for Strategic Studies.

Polonsky, A. (1975). *The Little Dictators: the History of Eastern Europe since 1918*. London: Kegan Paul.

Ramon, E. A. (1985). *Spain*. Boulder, Colorado: Westview.

Rosecrance, R. (1987). 'Long Cycle Theory and International Relations', *International Organization*, **41**, 283–301.

Rothschild, J. (1981). *Ethno-Politics – A Conceptual Framework*. New York: Columbia University Press.

Russett, B. (1985). 'The Mysterious Case of Vanishing Hegemony Or Is Mark Twain Really Dead?', *International Organization*, **39**, 207–32.

Samuels, N. (1986). 'Dealing with the International Debt Issue', in A. Henrikson (ed.), *Negotiating World Order*. Wilmington, Delaware, Scholarly Resources, pp. 199–216.

Seiler, D. (1989). 'Peripheral Nationalism between Pluralism and Monism', *International Political Science*, **10**, 191–207.

Sharpe, L. J. (1989). 'Fragmentation and Territoriality in the European State System', *International Political Science Review*, **10**, 223–37.

Shiels, F. L. (1984). *Ethnic Separatism and World Politics*. Lanham, MD: The University Press of America, pp. 3–10.

Short, J. R. (1982). *An Introduction to Political Geography*, London: Routledge & Kegan Paul.

Small, M. and J. D. Singer (1982). *Resorts to Arms: International and Civil Wars, 1816–1980*. Beverly Hills Calif.: Sage Publications.

Smith, A. D. (1982). 'Nationalism, Ethnic Separatism and Intelligentsia', in C. Williams (ed.), *National Separatism*, Vancouver, University of British Columbia Press, pp. 17–41.

Smith, G. (1989). 'Gorbachev's Greatest Challenge: Perestroika and the National Question', *Political Geography Quarterly*, **8**, 7–20.

Smooha, S. (1978). *Israel, Pluralism and Conflict*. London: Routledge & Kegan Paul.

Snyder, L. L. (1982). *Global Mini-Nationalism: Autonomy or Independence.* Westport, Connecticut: Greenwood Press.

Strange, S. (1987). 'The Persistent Myth of Lost Hegemony', *International Organization,* **41**, 551–74.

Taylor, P. J. (1989). 'One Worldism', *Political Geography Quarterly,* **8**, 211–14.

Taylor, P. J. (1989). *Political Geography.* London: Longman (2nd edition).

Thompson, W. R. (1973). 'The Regional Subsystem: A Conceptual Explication and a Propositional Inventory', *International Studies Quarterly,* **17**, 89–117.

Thorburn, H. G. (1989). 'Introduction', *International Political Science Review,* **10**, 179–82.

Thrift, N. (1986). 'The Geography of International Economic Disorder', in R. J. Johnston and P. J. Taylor (eds), *A World in Crisis?.* Oxford: Basil Blackwell, pp. 12–67.

Tura Solé, J. (1989). 'The Spanish Case: Remarks on the General Theories of Nationalism', *International Political Science Review,* **10**, 183–9.

Urquhart, B. (1989). 'The United Nations System and the Future', *International Affairs,* **65**, 225–31.

Wallerstein, I. (1974). *Modern World Systems, Vol. I: Capitalist Agriculture and the Origin of the European World Economy in the Sixteenth Century.* New York: Academic Press.

Wallerstein, I. (1983). *Historical Capitalism.* London: Verso.

Walt, S. M. (1989). 'The Case of Finite Containment – Analyzing U.S. Grand Strategy', *International Security,* **14**, 5–49.

Waterman, S. (1990). 'Involuntary Incorporation – The Case of Israel', in M. Chisholm and D. M. Smith (eds), *Shared Space, Divided Space.* London: Unwin.

Wehr, P. and Washburn, M. (1976). *Peace and World Order Systems.* Beverly Hills, Calif.: Sage Publications.

Williams, C. H. (1982). *National Separatism.* Cardiff: University of Wales Press, pp. 1–5.

Williams, M. (1989). 'China and the World After Tian An Men', *World Today,* **45**, (8–9), 127–8.

Williams, P. (1989). 'U.S.-Soviet Relations Beyond the Cold War?', *International Affairs,* **65**, 273–88.

van der Wusten, H. (1985). 'The Geography of Conflicts Since 1945', in D. Pepper and A. Jenkins (eds), *The Geography of Peace and War.* Oxford: Basil Blackwell, pp. 13–28.

Vayrynen, R. (1984). 'Regional Conflict Formations: An Intractable Problem of International Relations', *Journal of Peace Research,* **21**, 337–59.

Yadlin, R. (1988). *Arrogant and Oppressive Genius.* Jerusalem: Shazar (Hebrew).

2 The Emerging World Map of Peace

Saul B. Cohen

Introduction

As power becomes more diffused across the globe, observers tend to draw two diametrically opposed conclusions about prospects for peace. One view is that the decline of superpower domination and the rise of economic interdependence means not only far less risk of global war, but presages a general condition of peace and prosperity. The other is that there is every reason to assume that the 1990s will be even more unstable, as clashes of economic interest supercede or exacerbate territorial or ideological disputes. Such a view holds that the diffusion of economic power provides more tinder for conflict as the superpowers lose their capacities to serve as policemen of the world. Some of the pessimism rests upon historical analogies that relate cycles of hegemonic growth and decline to changing economic processes that undermine an existing world economic system (Wallerstein, 1983; Modelski, 1983). The cycle is one of hegemonic power rise, maturity, challenge and fall in a burst of global conflict, to be replaced by the emergence of a new set of global powers.

We are, in fact, in a period of hegemonic power decline and international reorganization. The sequence of events, from a post-World War II bipolar division of the globe to multipolarity and then the rise of regional powers now followed by political upheaval in Eastern Europe and collapse of the Soviet-imposed order there, attests to the deconcentration of political power. But a theory that posits an inevitability to the breakup of the world system through global conflict, to be followed by a new cycle, is outmoded. The decline of hegemony, while punctured over the years by conflict, is nevertheless taking place without global conflagration. For one thing, the nuclear standoff has provided a 'soft landing' for the devolution of power. For another, the concept of the self-contained national state that seeks to enlarge its resource base at the expense of non-sovereign territory – the goal since Westphalia – suits the colonial world of the past, not one of today, where most of the world's land mass enjoys sovereign status. And finally, hierarchical geopolitical integration is not collapsing into some kind of systemic abyss; it is, instead, being replaced by a flexible system of integration.

If I am optimistic that the world is not degenerating into global conflict as the end of a cycle, mine is also not the optimism of those such as Francis Fukuyama, who believe that we are witnessing the end of history with triumph of liberal democracy over Marxism-Leninism (Fukuyama, 1989).

He, too, feels that the new era means decreased likelihood of large-scale conflict although terrorism and liberation wars will still rage. But to assert, as does Fukuyama, that we are entering an era of post-history without excitement and challenge – indeed a sad time – is to assume a halt to the continued evolution of the world system. This is a simplistic notion. First, the demise of totalitarian Marxism in Eastern Europe and the USSR does not mean the inevitability of its replacement by Western democratic liberalism. Market economies may well flourish within a genuine democratic socialist system. Even if this should not be the case, a declining American economic system and a failing social order in the big cities, a Japanese society that has yet to experience its political convulsions, or a West European world which may turn more inward with integration, are weak reeds on which to pin a theory of the lasting triumph of democratic liberalism. In addition, with the majority of the people of this world still living in economic and political systems that do not belong firmly to either of the industrialized world's major ideological camps, the end of history is clearly not in sight. What we can say from our vantage point of optimism is that the pace of the geopolitical development process is accelerating, and the economic reordering has yet to put its stamp fully on the international political order. That these are times of profound international change – in the USSR, East and West Europe, the Middle East, Central America, China, Southeast Asia, is a statement of the obvious. As world statesmen seek to navigate through this maze of international uncertainty, the hesitancy of many to take advantage of opportunities and structures that can enhance global stability can be ascribed, in part, to innate fears of chaos. However, it is not chaos, but change and reordering that is now being experienced, and it is a reordering that has been taking place since the 1960s. Understanding how and why geopolitical processes work, leads to the conclusion that our world is not spinning out of control. Rather, it is entering a newer and higher stage of development.

World system change and wars

The geopolitical theory that prompted us to be optimistic about the map of peace is rooted in general systems theory (Werner, 1957; Cohen, 1982). The world is a complex system of geopolitical realms, regions, national states and subnational units. As an evolving system that is hierarchically, but flexibly organized, the system can reduce conflict at the higher levels, the global and regional, and yet accept conflict at the lower orders – the local. Moreover, given an adequate time frame – and to look at the decade of the 1990s is not adequate – we can project that peace at the upper levels of the hierarchy will ultimately have the effect of reducing local wars, especially if the ranks of the lower orders of the hierachy are augmented by states that arise in response to a process of trade and peace.

The recent steps toward military and ideological accommodation taken by the major powers, and the dampening of regional wars, are based, of course, on economic as well as military and political trends. The global economic crisis of debt and trade imbalance, and economic stagnation in most of the Third World and the Eastern Bloc have put heavy pressure on nations to reduce armaments production or acquisition.

Neither the US nor the Soviet Union are able to bankroll conflict elsewhere in the world with the same lavish hand as in the past. This is putting a damper on the continuation of costly regional and local wars. Thus, O'Loughlin's study of superpower competition and conflict in the Horn of Africa, Afghanistan and Central America (1986), while historically accurate, analyses an era that seems to be coming to an end as both the United States and the USSR experience economic crises that overwhelm ideological interests. Moreover, OPEC states now struggling with major treasury shortfalls can no longer provide their clients with the blank checks to pursue campaigns of war and terror.

Even if major and regional powers are less likely to war among themselves, and less able to stir the flames of conflict elsewhere, it is still necessary to include local war in the geopolitical stability equation. There have been nearly 150 wars since 1945, and from 20 to 30 still rage. The actual figure depends upon the measurement criteria employed. World Priorities, Inc., using a minimum of 1,000 deaths per annum, counted 22 wars in 1987, with a toll of 2.2 million lives, including 1.8 millions from war-related famine. The Stockholm Peace Research Institute (SIPRI) counted more than 30 wars in that year (Gartner, 1988).

Local wars can be very costly in human lives – witness the loss of half a million people in both Ethiopia and Mozambique during the past decade, or the uncounted thousands of lives lost in the Sudan and Angola.

Superpowers and regional states cannot, even if they truly desired, eliminate all local wars. But, they can help to minimize their intensity and duration by restricting the flow of arms and other forms of external support. For the most part, however, only total victory by one side or mutual exhaustion can end local wars. It is at this last stage of such conflict, the negotiations over peace, that external power-intervention has its strongest impact potential.

Local wars are themselves the result of deep-rooted domestic and regional ethnic, religious, racial or economic tensions, often latent until unleashed by the emergence of new states. Thus, the proliferation of national states in the world system has been accompanied by geopolitical instability – an instability born of conflict not subject to the nuclear deterrent that has maintained the uneasy great powers balance of the post-World War II era.

Future consequences of state proliferation

In recent history, state proliferation has been rapid. The number of sovereign states in the world, 175, represents a quadrupling since 1939. The attending conflict has either been part of the birth-pangs of the state, or it has come after independence as part of the struggle for political dominance by rival groups, or for extending or securing territorial claims. In other words, new states have been both cause and effect in the process of conflict.

If state proliferation has been a global destabilizing force, two important questions logically follow: (1) will such proliferation continue, and if so, (2) will this inevitably cause continued disequilibrium? My response to the first question is 'yes', and to the second, 'no'.

I expect proliferation to continue over the next two decades, albeit at a slower pace. However, many of the new states will be established as economically specialized 'Gateway' states, not subsistence-based territorial entities. Small states with limited sovereignty, these Gateways will emerge in a peaceful context, out of a process of negotiation and accommodation, because most of them are not strong enough to emerge out of revolutionary upheavals: by becoming links in our increasingly interdependent world, they will serve the cause of peace. Of course, new states will also emerge out of current violent struggles for territorial independence, but they will be fewer in number and carry less weight in the accommodation-conflict dichotomy.

Many of the new Gateway states will be located along the border zones of the world's geostrategic and geopolitical regions. They will thus help to convert these zones from zones of conflict into zones of accommodation. While in the early post-World War II era, the geostrategic border reaches, in particular, were flash-points for major power conflict, this type of conflict has now become muted. Outside of border zones, Gateway states will also emerge as trading states within regions that are experiencing rapid integration, like the European community.

Rather than along geostrategic borders, the greater threat to peace has recently been, and will continue to be, in the world's three Shatterbelt regions – Southeast Asia, the Middle East and sub-Saharan Africa – where local wars have raged unabated since 1945. However, many of these wars are winding down because of the combination of larger power disengagement from localized conflicts and mutual exhaustion of the antagonists.

First, let us examine the assumption that there will be continued growth in the number of states constituting the world map. The theoretical basis for the assumption is that the international system of states behaves in the same way that every general system behaves. As the system matures, it becomes more specialized and integrated, the number of its parts increases and the exchanges among these parts flow in freer, more flexible pathways. Applying this theory to national states, we observe that after the post-World War II breakup of colonial empires, a relatively large number of small states emerged. Each was only modestly differentiated from the others. These states copied one another by seeking to create relatively self-sufficient economies. They tried to close off their borders for security and nationalistic purposes. Western and Central Europe between the world wars were examples of this condition, but in fact this was the overriding characteristic of the sovereign territorial state that emerged after the treaty of Westphalia in 1648.

The nineteenth-century European colonial powers continued to seek such a protected, self-contained enclosed space system by extending their rule into colonial realms. Elsewhere, after World War II, new and competing states were established in the former colonial world. Where, instead of territorial states, federations or other kinds of mega-states were tried, with the most prominent exception of India and Nigeria which are sorely tested by conflict, they failed to endure – in the Caribbean, in Africa, in the Middle East and in Asia.

For many of the new post-World War II unitary states, the early differentiation has very quickly evolved into specialization – in agriculture and industry, resulting in greater trade in raw materials and finished goods. This development process has also evolved into greater specialization through other kinds of exchanges – in communications, capital flows, education, and immigration and tourism. On a regional scale, the European community, emerging from the Marshall Plan and the first phases of the common market, is an excellent example.

A more specialized integrated world system gives prominence to particular geographical areas within national states that serve as the links within the global system. This system presents opportunities for such areas to acquire some form of political independence, qualified as it may be, as a way of enabling them to maximize their specialized roles, for the benefits of others as well as themselves. The more integrated the world becomes, the greater will there be a need for certain portions of existing states to gain the flexibility of interaction that can only come through some form of sovereignty, qualified as it may be. The ideally advanced general system has countless numbers of parts or hinges that can connect with each other without having to move through rigidly controlled and hierarchical pathways. This is a model for the international state system too.

Now let us turn to the practical grounds for anticipating continued state proliferation. There are still many distinctive populations in clearly defined territories which would be candidates for independent statehood if the benefits were seen by their current sovereign governments as outweighing the costs. This includes about 30 political units that are in some form of dependency (colonies, dependencies, external territories). Examples are Bermuda, the Falklands, and Christmas Island. There are also twenty unit areas that are trust and self-governing territories, protectorates, departments or commonwealths. A few of these are the US Pacific Territories, Martinique and Guadeloupe, and Puerto Rico.

Because most of these are small islands, their restricted economic bases and physical isolation from one another decreases the conflict probability index. For strategic reasons, major powers will hold on to some (e.g. Diego Garcia), but for economic and political reasons, they will probably withdraw from many others. In addition, there remain a significant number of separatist or irredentist groups committed to military or political struggles for freedom. There are more than 50 units in this group. Repeatedly in the news are the Eritreans, the Southern Sudanese, the Polisario, the Northern Irish, the Kurds, the Tamils, the Kachins and the Palestinians. For many in this group, conflict is likely to persist, and will probably become more devastating as technological change permits the use of small-scale, but increasingly lethal weapons. On the other hand, resolution of the conflict seems at hand in the Western Sahara, and, despite the last-minute eruption of fighting, in Namibia. Within the foreseeable future New Caledonia, the Sudan, and the West Bank of Palestine may also find their ways to independence as host states experience political, psychological or military/economic exhaustion. There is little to suggest, however, that independence struggles in East Timor or Western New Guinea, in Nagaland or in Kurdistan will succeed within this generation.

Finally, there is a substantial number of peoples with unique cultural and historical bases who, while not yet caught up in violent separatist or irredentist struggles, might find it possible to fulfil long-held desires for self-determination because of unique economic circumstances. I refer to the pressures upon many national states to promote certain border areas as zones or territories for specialized manufacturing, tourist and financial services. Again, there are as many as 50 such units which could become candidates for independence, from Estonia and Slovenia to Madeira. Should the native Taiwanese gain control of Taiwan from the dominant nationalist minority, they might then secure agreement from China for their exercising a unique, if military subordinate, form of independence, and would join this category of states. These economic circumstances might also ease the way to independence for groups already caught up in violent struggle, such as the Basques.

What kinds of new states?

This discussion makes the assumption that nearly all of the states that will emerge as new members of the international system will be mini- or micro-states. This has been the case for most of the newly established units of the past decade. For such states, neither self-sufficiency nor self-defence is a realistic goal. Sovereign foreign policy for them is above all sovereign economic and cultural policy, not a political alternative to the military promotion of ideological goals. I want to emphasize that cultural homogeneity alone is not a sufficient criterion for the state that I have in mind – it must have a favourable economic base so that it can take advantage of the freedom to conduct its economic affairs. In some situations, new nations may go as far as did Bhutan when it agreed to be guided by the government of India in its external affairs. Though these two countries are now locked in a bitter dispute over transit rights and trade arrangements in their negotiations to renew the arrangement, there is little to suggest that it will not be renewed. Conventions between France and Monaco place the principality under France's protection. Independent former British colonies in the Caribbean, while having organized police forces, are dependent on the UK for defence, as is the case for six former French colonies in Africa where 8,000 French troops are stationed in their countries to guarantee internal stability as well as protection from external aggression – e.g. the Central African Republic, Chad, Djibouti, Gabon, the Ivory Coast and Senegal.

Dependence of independent states on intervening support from outside military forces (e.g. Lebanon on the Syrians, or Sri Lanka and the Seychelles on India) foreshadows additional formalized arrangements whereby small states will turn over their defence burdens to outsiders.

State proliferation has flourished and will continue to do so because the very small state with limited sovereignty probably has as much of a future as do the two basic models of the national state that have characterized the modern traditional territorial state, the ideal homogeneous, self-contained nation and the federated, heterogeneous state. The epitome of the homogenous state which aspires to cultural and economic territorial self-sufficiency protected by natural boundaries is France. Yet events long ago have shown that France and those national states that sought to emulate her could not go it alone.

The other ideal – the large, heterogeneous state that would achieve its goal of self-sufficiency through some form of federation or confederation is also far from a successful model, witness Nigeria, Yugoslavia, the USSR or India – or federations that emerged still-born, like the West Indies Federation, or that quickly fell apart, like the United Arab Republic.

Small states, with qualified sovereignty and specialized functions, that have recently joined the ranks of the international community do not have to apologize for falling short of the criteria set by the failed idea state models. Because they fall short structurally, they may be derided as 'pseudo-states'. But who is to say that their functions within the international community are of lesser worth than that of homogeneous national and the federated state models? I believe that the new type of small state, the Gateway, is about to join the current batch of ministates. Rosecrance, in writing of the distinction between trading state system and the territorial state system, speaks of the role of the 'mediative state' (Rosecrance, 1986). This refers to a state that is an intermediary between internal (territorial sovereignty) and international (trade-dependent) pressures.

While many states are mediative in the sense that they contain within them and are therefore caught between such pressures – this includes the US and the USSR – in some cases it is difficult to resolve such contradictions within a sovereign territorial framework. The optimal solution for mediative host states may not be to try to balance off their internal contradictions, but to agree to the detachment of certain parts to form separate Gateway or 'exchange' territories that can act with the relative freedom that limited sovereignty provides.

Perhaps the systems analogy here is to microelectronic circuitry, or chips, in which gates permit currents to pass through arrays or transistors. Transistors are made faster by making them smaller, giving the current less distance to traverse. Some Gateway states already exist – Finland, Bahrain and Malta. Lebanon was also a Gateway before its *de facto* dismemberment 15 years ago. These states are not simply colonial-type transhipment centres, like Djibouti. They have national cohesion and an economy that is a substantial value-added centre for material, peoples and ideas. Finland, for example, has a free-market economy that relies on Soviet trade, but is open to the West, is neutral in foreign affairs and free of Soviet military bases, but attentive to Soviet security needs including a requirement to join the USSR in repelling any attacks on the Soviet Union from the West.

The locations of new states

Analysis of the map of the world shows us that most of the world's states created since World War II are located either in the world's three Shatterbelt regions – the Middle East, Southeast Asia, or sub-Saharan Africa – or in the Caribbean and Pacific basins. The Shatterbelt states especially are the ones that have been caught up in conflict – regional and local wars, and internal rebellion. In fact, the highest incidence and intensity of war since World War II has occurred in the Shatterbelts by a 2½ to 1 frequency, as van der Wusten and Nierop (1988) have affirmed in their studies.

Previously, I have suggested that the world consists of separate, hierarchically organized interlocking, geopolitical areas (Cohen, 1963, 1973). This world model is divided into two geostrategic realms, and 11 geopolitical regions, including three Shatterbelts. Of the geostrategic regions, the Eurasian Continental World is dominated by the USSR from its east European-inner Asian heartland core. This is the centre or pivot land of Eurasia in Halford Mackinder's terms (Mackinder, 1904, 1919). The other, the trade-dependent maritime world, has Anglo-America as its centre, and is linked by sea to the rest of the Atlantic and Pacific Rim realms.

Within these two geostrategic realms are smaller regions, organized as geographic, economic, cultural and military tactical frameworks around core states. Each of the trade-dependent maritime world's major powers – the US (Anglo-America and the Caribbean), the European Community (maritime Europe and the Maghreb) and Japan (Offshore Asia and Oceania), anchors a geopolitical region. While some would place Japan in an East Asian or Pacific basin region, I believe that the trade zone from Seoul to Sydney, to Singapore and Jakarta, which is held together by the yen, is a more valid geopolitical region than those frameworks that would

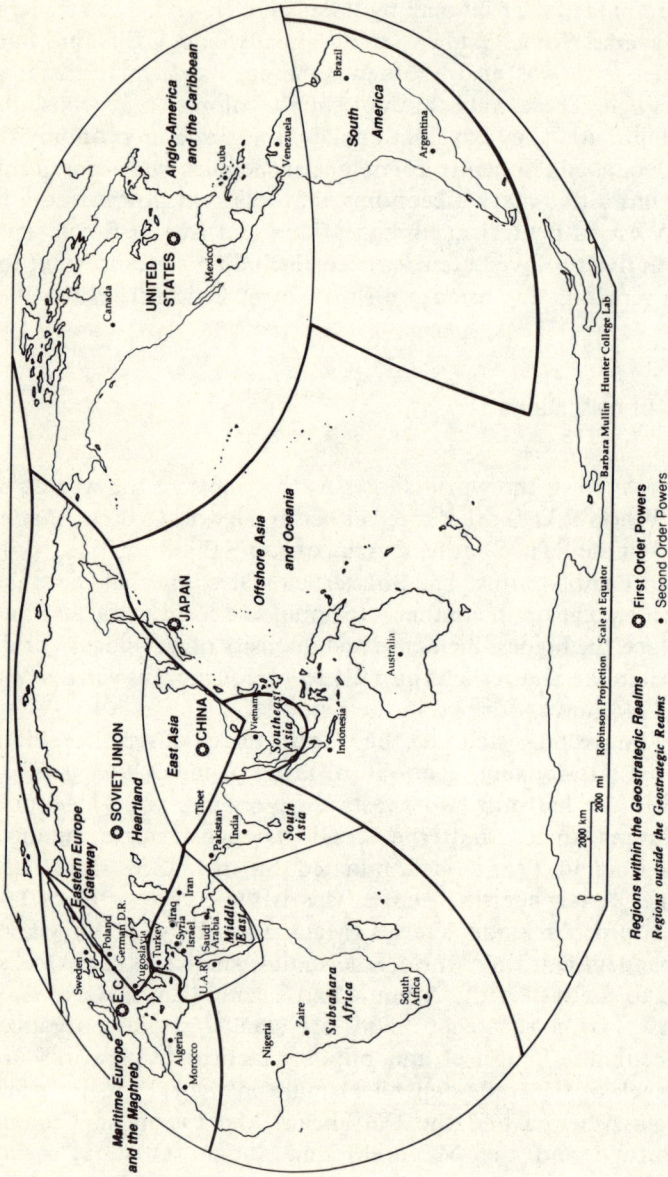

Figure 2.1   Geopolitical regions of the world

infringe upon either Chinese or US military political and ideological regional primacy.

Within their geostrategic realm, the USSR and China dominate their respective geopolitical regions. Other geopolitical regions are South America and South Asia. South America is part of the trade-dependent maritime world. India is the primary military and political power within the independent South Asia region. This power will increase as the superpower struggle over Afghanistan winds down and Pakistan's strategic value to the US is diminished. Finally, the three Shatterbelts complete the world's geopolitical map. The system that is suggested is regionally framed, with second-order powers dominant in those regions that lack a major power. All told, I have identified five orders of power. The third level, as well as second-order (regional) and first-order (major), states has the capacity for outreach to other geopolitical regions (Cohen, 1982). As the world system has developed, its regional components have become more open. The most prominent example of such opening is now occurring in Eastern Europe. The regional boundary between East and maritime Europe has become porous. Some Western specialists, especially the US, are beginning to express the fear that too rapid a movement toward political pluralism and economic reform may lead to geopolitical instability. It is ironic that such expression often comes from those who have been most vocal in their opposition to Yalta and the Iron Curtain.

In fact, as a system matures, it can be expected to become more open and responsive to change. Precisely because Soviet military might remains unchallenged in Eastern Europe and is not dependent upon any gratuitous promises from the United States 'not to rock the boat', the USSR can afford to encourage perestroika on the western edge of its geopolitical region. Those who see the crumbling of the Soviet Empire should keep Soviet military power in perspective. Withdrawal from Afghanistan was accomplished with dispatch, order and minimum loss. This is in contrast to the disorganized, precipitous American withdrawal from Vietnam, which was the humiliating denouement of the Nixon–Kissinger four-year search to save face in extricating the United States from the Vietnamese morass.

On the contrary, a self-confident Soviet Union which sees itself as the military equal of the United States can now permit greater flexibility and openess within its geopolitical region, than it could during the formative years of Comecon and the Warsaw Pact. This is another lesson to be learned from our understanding of systems development theory.

The boundaries of the various geopolitical regions may shift. For example, Indonesia and Singapore are now part of Offshore Asia, not Southeast Asia (Cohen, 1984). Also, the status of a region may be completely altered. When I first offered this view of the world, sub-Saharan Africa was part of the trade-dependent maritime world. Now sub-Saharan Africa is a Shatterbelt as a result of the forceful intervention by the USSR and its allies in such countries as Ethiopia, Mozambique and Angola, while

France and other European nations retain a strong military and economic presence among their client states, and the United States maintains its influence in the region, too (Cohen, 1982). Sub-Saharan Africa, like the Middle East, is a region whose external military and diplomatic ties outweigh its internal links, and whose international trade is overwhelmingly outside the region.

Shatterbelts are not fixed in space and change in their outlines or overall status, but the change is developmental. East and Central Europe was a Shatterbelt between World War I and II. It became part of a Soviet-dominated geopolitical region when the USSR rose to continental pre-eminence after Germany's defeat. Today, with the economic and political weakening of the Soviet Union and the emergence of a powerful integrated maritime Europe, the lands between the Elbe River and the Soviet border are on the verge of acquiring a new status.

Eastern Europe may be detached from the Soviet geopolitical region to emerge as a new kind of Gateway region. It most likely will remain under the military umbrella of the USSR. However, its economy will become fully open to both the West and the East, and its national cultures and political systems will achieve independence. The developmental sequence is from Shatterbelt to subordinate or peripheral part of a geopolitical region to a new kind of intermediate or Gateway region.

It is premature to suggest that the present three Shatterbelts of the world are on the brink of similar change of status. Until their intraregional conflicts are substantially reduced, change in status is unlikely. However, we can anticipate a stage when Southeast Asia will experience profound change, with Thailand and Malaysia joining Offshore Asia, Burma becoming part of South Asia, and Indo-China, with a renewal of positive relations with China, becoming part of the East Asia geopolitical region.

In the three Shatterbelts which have experienced the highest intensity and frequency of war in the past four decades, 56 out of the 68 states are new, having gained independence since World War II. The number of additional new states that may be created in these regions is now limited. Now that Namibia has attained independence, the most likely candidates for state-hood in the Shatterbelts in the foreseeable future are Gaza and/or the West Bank, Eritrea, perhaps Christian Mount Lebanon and Southern Sudan. So it is not to the Shatterbelts that we look for the emergence of many new states. Instead, we expect them to arise along the boundaries of geostrategic realms and geopolitical regions: on Pacific and Atlantic islands, and in regions where strong, macro-regional forces are making their impact – like the European Community and Anglo-America.

The borderlands between the various geopolitical regions are, increasingly, zones of accommodation, as the era of Containment and Iron Curtains has passed into history. In general, since World War II, the incidence of conflict along geopolitical borders has not been especially higher than elsewhere in the world. However, the nature of the conflict that has taken place has been fraught with peril to global stability, because the world's major powers became directly involved.

In the years following World War II, the boundary between the world's two great geostrategic regions was the world's most unstable conflict zone – from Greece to the Koreas, to the Chinese–India borderland, to Vietnam and Afghanistan. Now, war along the geostrategic border zone has largely abated. With the final withdrawal of Soviet armed forces from Afghanistan, foreshadowing a political or military resolution of conflict, the zone will be quiet. Major powers fear the escalation consequences of direct military confrontation with one another. So, overall, the geostrategic borderlands are beginning to shift from military flash-points to those in which normal economic relationships are beginning to take form. Developmentally, the shift is from boundaries as barriers, to the boundaries of accommodation that Lionel Lyde foresaw so many decades ago (Lyde, 1926).

It is in these zones of accommodation and contact, therefore, that we look for the emergence of most of the new Gateway states. The following table suggests some possible candidates.

Example of Gateway states

Among the states listed in Table 2.1, I shall comment briefly on three: Estonia, Yugoslavia and Madeira. It is quite clear that Estonia cannot gain sovereignty by force of arms. Moreover, whatever sovereignty is secured is unlikely to carry with it independence as well as cultural freedom and the ability to control immigration and thus preserve Estonian ethnic control may be a price that the Russians will ultimately be willing to pay. Such willingness would be based on self-interest, the prospect of a positive impact on the Soviet economy that a Baltic state can set through its economic flexibility and innovation.

Perestroika has many meanings, one of which is structural change that opens the Soviet system economically. Gateways can be part of such change, indeed an essential ingredient to Soviet economic growth. An Estonian Gateway state can perform a positive regional role, provided that its nationalist goals do not challenge Soviet security needs. An Estonian state that can preserve the 60 per cent ethnic mix, including its Lutheran traditions, has currency that can be freely traded with a West European common currency should it emerge, and can organize itself with its Baltic neighbours as a coordinated economic free zone, would be as helpful to the USSR as it would be to Estonia. There are, of course, limits to how far the USSR can go in responding to separatist movements. The Ukraine, Georgia, Armenia and Azerbaijan are parts, which, if lost, would dismember the whole. This is not the case for the Baltic republics.

A similar case can be made in Yugoslavia. Serbian assertiveness and Albanian Muslim nationalism in Kosovo has been described by some as the

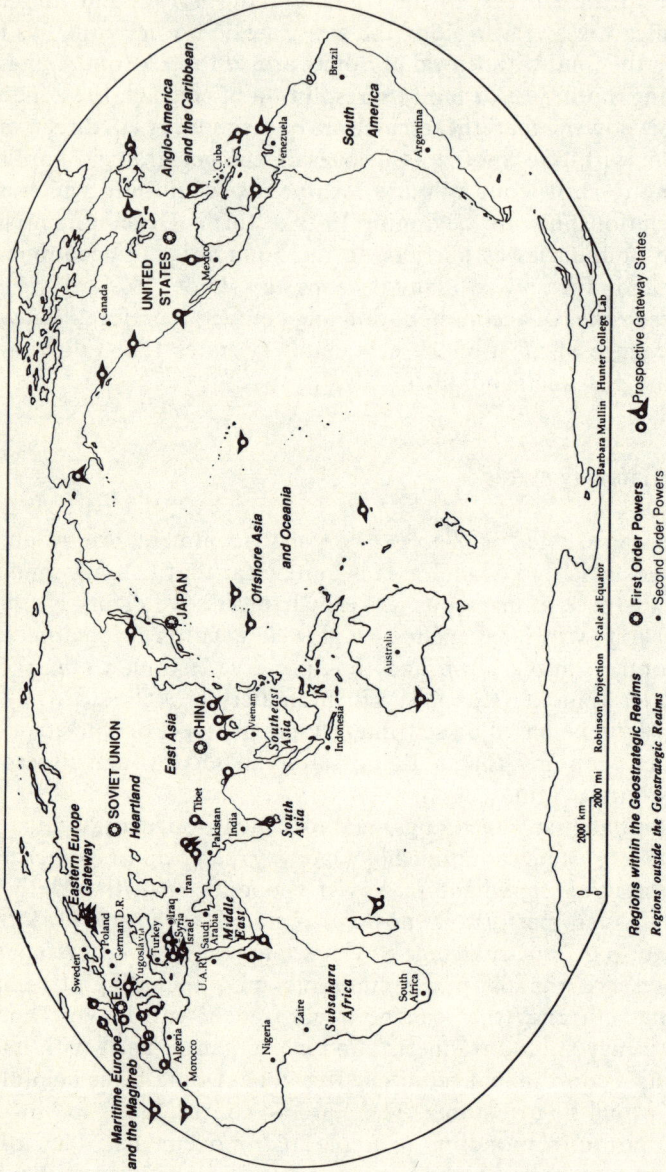

Figure 2.2   World Gateway states

Table 2.1   Prospective Gateway states

Inter-geostrategic realm borders Gateway state
Maritime Europe-heartland | Estonia, Latvia, Lithuania, Slovenia,
Heartland-East Asia– | Soviet Far East
  Offshore Asia
Heartland-Anglo-America/ | Alaska
  Caribbean
East Asia-Offshore Asia | Hong Kong, Macau, Coastal Kwangtung

Inter-geopolitical regional borders Gateway state
East Asia-South Asia | Tibet, Kashmir
East Asia-Southeast Asia | North Burma
  South Asia
Offshore Asia-Southeast Asia | Western Australia
South Asia–Middle East | Pashtunistan
South Asia–Sub-Saharan Africa | Reunion
Middle East–Maritime Europe | North Cyprus
Middle East–Sub-Saharan Africa | Eritrea, South Sudan
Maritime Europe–Anglo-America | Azores, Madeira, Canaries, Bermuda, Newfoundland
Anglo-America/Caribbean–South | Puerto Rico
  America
Anglo-America/Offshore Asia | US Trust Territories, American Samoa, Guam, Hawaii

Intra-geopolitical regions
Anglo-America/Caribbean | Quebec, British Columbia, US Pacific Northwest, California, Nicaragua, North Mexico, Netherlands Antilles
Maritime Europe/Maghreb | Northern Ireland, Scotland, Basque Country, Gibraltar, Catalonia, Sardinia, Bolzano
Middle East | Gaza (West Bank), Mount Lebanon
South Asia | Tamil Eelam

first stage of a choice between a Serbian-controlled centralized state and the destruction of the federation. Kosovo, however, is a minor player in the question of Yugoslavia's future. The threat of Albanian irredentism seems minimal when one considers that Kosovo's alternative to belonging to Yugoslavia would be to become attached to Albania – Europe's most closed and backward state.

A more significant potential break-away area is Slovenia. A Gateway, quasi-independent Slovenia would have benefits for Yugoslavia and Europe as a whole. Slovenia is more Central European culturally, historically and geographically, and more advanced economically than its fellow

Yugoslavian republics. It can show the way for Yugoslavia through an independent political structure that would not pose a military threat to Yugoslavia, but would provide a breeding ground for economic experimentation that could benefit all. Another Gateway example is the island of Madeira, several hundred miles off the coast of Portugal and Morocco, and on a line with Casablanca. Madeira illustrates the case of islands ripe for independent status. It is presently an autonomous region within Portugal. It has home rule over its regional budget and tourist development. However, its dreams of developing as an offshore banking centre and free trade zone have long been delayed by central government bureaucracy in Lisbon. As a base for companies seeking to export to the European Economic Community after 1992, the Madeira free zone could be quite attractive.

Madeira needs Portugal for the entry that is provided into the European Community. And, it has no incentive to take on defence burdens. But an arrangement that provides Maderia with the independence to take economic advantage of its Atlantic basin location for exports to Europe, the United States and North Africa, could free Portugal from what is now an economic burden. Madeira could be a model for the development of Europe's poorer regions. In the Pacific, Hawaii is another example of a border entity with Gateway state potential. Hawaii is so rapidly becoming a part of the Pacific Rim in economic and cultural terms, that some sort of political separation, while remote at the present, need not be totally out of consideration.

These are examples of islands that have limited defence and political foreign policy concerns so that they can remain under the military umbrella of the countries to which they presently belong. They can succeed as microstates because they have the ability to become heavily involved in financial services, capital flows and tourism. Often they need to be able to control immigrants. And sometimes they are ideal places for assembling manufactured parts into finished products. Some of these political states have access to the capital and technical know-how of emigré populations who left crowded, agriculturally impoverished island bases, but retain emotional and familial ties.

In certain situations, the new states that we are describing will not only maintain ties to mother countries, but may also form loose economic 'leagues' among themselves. This harks back to the Hanse towns of the late Middle Ages (1300s). These mercantile centres were spread along the Baltic and North Seas from Visby to Bremen, and up such main rivers as the Rhine, Elbe, Oder and Vistula, to Cologne, Magdeburg, Breslau and Cracow. The Hanseatic League, while principally consisting of German cities, had agents and affiliates from England and Bruges to Novgorod. The League's purpose, mutual relations and trade, could serve as a valid model today for linking small states that have the drive and flexibility to become centres of innovation today.

Regionalism and new states

While most Gateway states will emerge along the borders of geostrategic realms and geopolitical regions, some will arise within rapidly integrating geopolitical regions. The European Community is the most advanced. One might argue that the European Community is itself in the process of becoming a mega-state, and that we should anticipate the merging rather than the proliferation of national states in this region.

The evidence, however, is to the contrary. The European community is built on the framework of national states, many quite small. Its operational goals are to respect and nurture the specialized national qualities of its member states or sections thereof. The European Community is expected to complement the social, cultural, education resources and political needs of its members, not to undermine the national cohesion. While regional corporate, banking and research centres will thrive, as will agricultural specialization, there is little to suggest that national histories, cultures and languages will be seriously undermined.

The European Community presents an opportunity for European peoples who have sought independence – the Basques, Catalans, Northern Irish Catholics, Sardinians, Scots, Walloons, and Tyrolese (Wallonia may already have attained its desired status in Belgium's advanced confederal structure.) These smaller groups could survive economically in a Europe without meaningful national political boundaries. The detachment of these ethnic or religious minorities from the mother country would create no security problems in a Europe with a unified defence posture. Moreover, many of the economic advantages the mother country enjoys from having these subregions within their economic borders, could well be lost as the European Community enters into the new era of 'pooled sovereignty' in 1992 (Wise, this volume, Chapter 8) – or conversely the burden for their support could be shared.

Elsewhere, another major regional free trade and political entity is emerging in North America. The Free Trade Agreement between the US and Canada could be the forerunner of a broader one that includes Mexico, despite statements by Mexican leadership of their fear of such an eventuality. As with the European splinter groups, the US – Canadian Agreement could make it easier for Quebec separation to succeed.

Indeed, Jacques Parizeau, leader of the Parti Quebecois that is pledged to separate the province from the rest of Canada, feels that the Agreement will advance his cause (*New York Times*, 1988). Mr. Parizeau's point is that the US market, which now consumes only 30 per cent of Quebec's production, could become a much greater factor in the Quebec economy, freeing Quebec from the threat of boycott by the rest of Canada. As it is, Canada has negotiated an accord with its provinces that, if ratified, would recognize Quebec as a 'distinct' society. Such recognition, plus the trade agreement, could bring qualified independence much closer. British Columbia and the

United States Pacific Northwest are other examples of potential border entities.

US–Mexico relations are following pathways of mutual assistance and relatively free exchange of goods and people. The US needs guaranteed access to energy resources as well as Mexican labour, while Mexico's financial requirements are overwhelming. One might also foresee a time when Mexico's northern regions would seek some form of independence. In the Macquilladores zone, parts from the US are assembled and returned to the US duty free. This could conceivably be the core for such a new, autonomous national state. Finally, the emergence of a Pacific economic realm that might not only include Canada and Mexico, but also Japan, could have unforeseen geopolitical outcomes. A leading Japanese businessman has recently proposed that California become a US–Japanese condominium, with Japanese labour being allowed to migrate freely to California to relieve Japan's population pressure in return for Japan's taking the fiscal steps to balance its trade with the United States. While such a radical restructuring is unlikely, it is possible to foresee California enjoying a unique economic status within the United States, as it forges a new relationship with Japan. This might also be a scenario for the US Pacific Northwest and for British Columbia, which is experiencing an unexpected inflow of Asian capital and immigrants.

Conclusion

Some new states will continue to emerge in a context of conflict. However, most of the proliferation that is envisaged will emerge from and enhance the peace process. The peoples and territories to which we refer are already forging complementary ties with affinity sectors outside their countries. These areas are Gateways or links to the rest of the world, and potentially important new nodes in the international system. Such new states offer promise as centres of innovation, communication, capital generation and formation, and economic initiative. They can help change the economies of their current parent countries, if given the necessary independence of action.

We are witnessing the flowering of a new era in the life history of the international system. In the past century, new nation states emerged with the goals of broadening and diversifying their economics. They sought to become self-sufficient in wheat or steel, in petrochemicals or transportation equipment. In this sense, they made the world system more undifferentiated, stunting its development. Today, nations irrespective of their ideological cast, have come to appreciate the benefit of specialization. As states engage in this era of interdependence, the need for larger national territorial units decreases. The multi- and transnational corporation has substituted for this need and is, indeed, the pioneer of the new system. Irrespective of size, states can benefit from being integral parts of a more highly developed and sophisticated system, as an understanding of developmental systems theory teaches us. This is an era, then, that holds considerable promise for global stability, if the international system can evolve as a complex mosaic of states (themselves with differing functions), regions and realms, tied together by flexible, not rigid, hierarchical structures.

The analogy is to flexible automated systems that allow parts to be exchanged within a manufacturing assembly system in any sequence, without having to follow a linear assembly line. A flexible, international system would permit states to link up globally, regionally and sectorally.

The emergence of additonal, very special types of small states linked among themselves and to the world in manufacturing, technology transfer, tourism and capital flows, reflects a geography of affinity. They can contribute to a world that is pluralistic, polycentric and flexibly hierarchical in scale. The more highly developed the system becomes, the more it is able to cope with shocks, as blockage points are bypassed and the system feeds upon a multiplicity of nodes.

As the world geopolitical system rises to higher stages of development, becoming more specialized and integrated in a geopolitical structure that evinces greater degrees of hierarchical flexibility, the parts of this system will continue to multiply. Whereas, in the recent past most new parts emerged from the breeder process of decolonization, in the future territories of large states may become the direct spawning grounds of many of the world's new states. If territorial detachment leaves the core of the large state intact, then it can afford to permit such a breeder process to occur.

Thus, we can anticipate a national state proliferation that will contribute to the geography of the peace. It is in this context that the Gateway state, in particular, will shape the map of the world to come and contribute to the stability of the global geopolitical system.

References

Atlas, J. (1989) What is Fukuyama Saying? *The New York Times Magazine*, 22 Oct. 1989, Section 6, 38–42, 54–5.

Cohen, S. B. (1963) *Geography and Politics in a World Divided*. New York: Random House.

Cohen, S. B. (1973) *Geography and Politics in a World Divided* (revised 2nd edition). New York: Oxford U. Press.

Cohen, S. B. (1982) A New Map of Global Geopolitical Equilibrium: Developmental Approach, *Political Geography Quarterly,* **2**, 223–41.

Cohen, S. B. (1984) Asymmetrical States and World Geopolitical Equilibrium, *SAIS Review,* **4**, 193–212.

Fukuyama, F. (1989) The End of History? *The National Interest*, Summer, 1989.

Gartner, M. (1988) A Dream of Peace, The Reality of War, *The Wall Street Journal*, 22 Dec, 15.

Lyde, L., (1926) *The Continent of Europe*. London: Macmillan & Co. Ltd.

Mackinder, H. J. (1904) The Geographical Pivot of History, *Geographical Journal,* **23**, 421–37.

Mackinder, H. J. (1919) *Democratic Ideals and Reality.* New York: Hold (reissued W. W. Norton, 1962).

Modelski, G. (1983) Long Cycles of World Leadership. In: W. R. Thompson (ed.), *Contending Approaches to World System Analysis.* Beverly Hills Calif.: Sage Publications, pp. 115–39.

The New York Times, Sunday, 4 December 1988, 16.

O'Loughlin, J. (1986) World-Power Competition and Local Conflict in the World. In: R. J. Johnston and P. J. Taylor (eds), *A World in Crisis: Geographical Perspectives.* Oxford: Basic Blackwell, 231–68.

Rosecrance, R. (1986) *The Rise of the Trading State.* New York: Basic Books.

van der Wusten, H. and Nierop, T. (1988) 'Functions, Roles and Form in International Relations', *Mimeo.*

Wallerstein, I. (1983) *Historical Capitalism.* London: Verso.

Werner, H. (1957) The Concept of Development from a Comparative and Organismic Point of View. In D. Harris (ed.), *The Concept of Development.* Minneapolis: Minnesota University Press, pp. 125–49.

3 From 'Geopolitik' to 'Geopolitique': Converting a Discipline for War to a Discipline for Peace

John O'Loughlin and Henning Heske

One of our most important and long-standing traditions in political geography is undergoing a belated renaissance and at the same time, a rigorous scrutiny of its theories and applications. We refer to the political geography of international relations, better known as geopolitics. The field is under-theorized, under-researched and still dominated by the geographic determinist model (O'Loughlin, 1988). A return to basics is needed including a reconsideration of the roots of the geopolitical tradition in geography and the formation of a new approach to the geography of international relations that might be termed a 'new geopolitics'. (Agnew and Corbridge, 1989). Consequently, in this chapter, our aims are two-fold; to bring the 'imperialist' legacy of political geography to the foreground and to reclaim the tradition of geopolitics from the strategic community, by demonstrating the false nature of their assumptions and prescriptions.

The title of this chapter reflects our belief that the strategic vision still dominates geopolitics. The journal *Geopolitique*, published in Paris and Washington DC, continues the hegemony of the 'false god' of geographical determinism, relegating the political geography of international relations (IR) to a role as successor to the German school of Geopolitik. Within the realist (power-political) framework of IR, the geographical contribution is reduced to the strategies of territorial control and the apposition of land to sea-power. (See the chapters in the NATO publication of Zoppo and Zorgbibe, 1985.) We agree with Lacoste (1976) that 'La geographie, ca sert, d'abord, a faire la guerre.' In academic writings and in the public expression of the discipline, geography of the late nineteenth and early twentieth centuries was the servant and promoter of imperialism and territorial expansion. There were three lonely protesters in the discipline (Reclus, Kropotkin and Wittfogel), all of whom were hounded out of their home-lands. After 1945, geopolitics became the preserve of the American strategic community, having fallen out of favour in the decolonizing European states. It is time to reclaim the geopolitical theme from its hijackers in the strategic community.

The geopolitical literature can first be divided into Kleingeopolitik (the geopolitics of an individual state) and Grossgeopolitik (global-level views and strategies). Toal (1989) classifies strategic geopolitics into six genres, the

classical nineteenth-century 'Great Power' strategies of Mahan and Mackinder, German Geopolitik, post-World War II American geopolitics, a
'depoliticized' tradition that tried to identify the role of geography in IR
(Cohen, 1982), policy-oriented strategic studies of contemporary American
writers, and a national security state (South American) geopolitics.
Together with the Lacoste school, taking its legacy from Reclus (see below),
this is a classification that rings accurate to us. What is most interesting and
important is that it is the American strategic school that adheres most
closely to the tenets and beliefs of the classical school, especially in the
belief, following Spykman (1938: 29), that geography is 'the most
fundamental conditioning factor in the formulation of national policy
because it is the most important'. While political geographers try to
incorporate consideration of social, cultural, economic and political processes into their researches, the strategists in the policy world are,
perversely, the most determinedly geographic.

Geography and imperialism, 1870-1918

Geography as a university discipline emerged as a result of imperialism
(Taylor, 1985). The Franco-Prussian War of 1870/71 was the political event
that led to the establishment of academic geography; it was 'a war fought as
much by maps as by weapons' (Hudson, 1977: 13). Indeed, among the
Prussian officers of 1870 were former students of Carl Ritter, who had
taught at the Military Academy as well as the University of Berlin. As a
reaction to defeat, France also encouraged the development of geographical
institutes and organizations (McKay, 1943). Geographers in both countries,
and also in Great Britain and Italy, used newly acquired national reputa-
tions for the promotion of colonial expansion. Geographers were
successively involved in territorial mapping, economic exploitation, military
actions and the practice of class and race domination. Geography, in effect,
became a tool of imperialism.

   In the early phase of imperialism, geographers took on the roles of
political initiators rather than those of consulting scientists. They advocated
especially the exploitation of Africa (Heske, 1987b) as geographers saw a
chance to bring their discipline to prominence. A second pact developed
with trade and industry, bringing financial support for geographical
explorations, organized and promoted by the various national geographic
societies. Nationalist and power political ideas, supplemented by social
Darwinist and racist attitudes, were all-pervasive and helped to justify for
domestic populations the warfare against natives as well as contending
colonial powers.

   The 'dream of Africa' (Der Traum von Afrika) was formed by myths
about treasures and tropical fertility, partly resulting from sensationalized
reports about the travels of European explorers. During the late nineteenth

century, the most interesting spot in the world for geographers was the heart of Africa, the Congo Basin. The Congo Conference of the colonial powers in 1884 in Berlin arranged the diplomatic rules for the 'scramble for Africa' and within a few years, the continent was completely partitioned among European powers. As a result, many ethnic groups were brutally separated into different territories (Krings, 1984).

During the European territorial expansion to all corners of the world, two early geopolitical theorists developed the 'pseudo-scientific' arguments with which such a policy could be justified. By his 'laws of the spatial growth of states', Friedrich Ratzel described the expansion of a state, through war, as a natural progressive tendency. 'The space of states grows with their culture' (Ratzel, 1896:31) and his social Darwinist approach is well-documented (Bassin, 1987). Ratzel also stated (1896:39) that the greatest successes of expansive politics is predicated on the use of geography. Subsequently, he subtitled a textbook on political geography, a work with clear implications for imperialism (Bassin, 1987), 'the geography of states, traffic and war' (Ratzel, 1897). Sir Halford Mackinder's (1904) theory about the global balance of power became another justification for imperialist policies. His model of an Eurasian continental bloc as the geographical pivot of history retained great influence on the foreign policy of the world powers until the end of World War II and today is used by American strategists to legitimate the containment strategy (Luttwak, 1983; Gray 1988, 1989; Brzezinski, 1986).

Geopolitical thinking from nationalistic and aggressive viewpoints produced much work in mapping new states and world orders. One of the most famous is the anonymously published book *Germania Triumphans*, which described, from a German point of view, the expected global conflict as early as 1895 (Sandner, 1989). In this utopian war scenario, after its assumed beginning in 1903, first allied Germany and Italy defeat France. Then Germany, Italy, Austria and Turkey beat Russia. The result was a new political map of Europe, incorporating an all-European customs union but excluding Britain. In a second phase of war beginning in 1912 the European powers defeat the USA in South America and subdivide Latin America. Finally a third phase in 1913 led to a European victory over Britain resulting in a redistribution of Africa and a re-division of the colonial spoils. This new political map is 'a reflection of the intentions in colonial and Pan-German circles at the end of the century' (Sandner 1989).

It is no secret that geographers were active members of imperialist circles. For example, in 1890, Friedrich Ratzel founded the 'Alldeutscher Verband' (PanGerman Union), a political party with enormous imperialistic aims, including the maintenance of the dream of a great German colonial empire in Central Africa, as envisaged in *Germania Triumphans*. Ratzel's student, Bruno Felix Hansch (1912) developed in a prestigious journal, *Geographische Zeitschrift*, a re-division of Africa with a great German-Belgium Central African empire. This economic region would include the Belgian Congo as an economic base and the German colonies and Angola as its periphery, in an area over 11 times larger than the territory of the German Reich. In this context, the term 'Lebensraum', which was promulgated by Ratzel, gained prominence although the term had already been used by German geographers more than twenty years earlier (Faber, 1982: 400). 'Lebensraum', (living space), a biogeographic term with clear social Darwinist implications (W. D. Smith, 1986), became one of the key arguments of the post-1918 German geopolitics of war, with German geographers and politicians claiming an expansion of the shrunken German territory as a result of the Treaty of Versailles.

The geopolitical paradigm changed during the nineteenth century from natural boundaries as contributing to the peace of nations to a Ratzelian paradigm of the struggle between expanding territories, with war as the natural mechanism in the struggle for space (Schultz, 1987). World War I subsequently marked the climax and final phase of this first geopolitical era.

The rise and fall of the geopolitics of war, 1919-45

The Treaty of Versailles in 1919 was a geopolitical act with far-reaching consequences. Germany lost large parts of its territory and all colonies: its world empire ambitions were abruptly and severely damaged. The American geographer, Isaiah Bowman, mapped this 'New World' (Bowman, 1921). Bowman, Director of the American Geographical Society, was Chief Territorial Specialist to the US delegation at the Paris Peace conference, where his maps were used by a number of the territorial commissions (Smith, 1984). Likewise, E. de Martonne advised Clemenceau, the French leader, on central European boundaries at the conference (Herodote, 1987: 7). After this conference Bowman helped found the Council on Foreign Relations in 1921 to promote an activist and globalist foreign policy for the United States. Bowman explained the conditions for an American economic expansion in his book *The New World* (1921), which contains overtones of Social Darwinism as well as a modicum of racial prejudice (N. Smith, 1986). Until the end of World War II, Bowman, still very influential in the US foreign policy establishment, considered war as an essential ingredient of power politics:

> Defense is a part of our way no matter through what seas of blood it leads – or we lose the way of life we cherish. The soldier on a Greek vase of the fifth century B.C. carries a sword without apology: to the Greeks, war was one of the arts (Bowman, 1942: 352).

As Bowman was developing an American geopolitical world view to cope with the new global distribution of power, General-Major Karl Haushofer and other German geopoliticians created 'Geopolitik' as a response to Anglo-Saxon (Mackinder and Bowman) global models. The main aim for Haushofer, who became Professor of Geography at Munich, and the contributors to his journal *Zeitschrift fur Geopolitik* (1924–44) was the reunification of all Germans in Europe in a great German Reich. This panGermanic approach was founded on the theory of 'Lebensraum', the key concept in German geopolitics after Ratzel (Heske, 1987a; Kost, 1988). Haushofer drew up a global model of panregions, which he conceived as political-economic unions of nation states, of sufficent size to discourage attack and to permit economic autarky (Heske and Wesche, 1988). Each panregion would have a developed, dominant core and a resource periphery. Haushofer considered a fourfold division of the world as the most likely scenario. Japan, possibly in conjunction with China, would dominate Pan-Asia including the periphery Australia. The United States would lead Pan-America. In the panregion Eurafrica, a unified Europe with a German core, would retain control over Africa and the Near East. The USSR, possibly with minor appendages in Asia, would likely constitute a fourth panregion. Haushofer held out the hope that panregions would evolve through accommodation between states in recognition of geopolitical realities and mutual interests, but he did not exclude the possible use of force to achieve this global equilibrium (O'Loughlin and van der Wusten, 1990). The concept of a super continent, Eurafrica, was sanctioned at the same time by German colonial geographers like Erich Obst, who had already postulated a return of lost German colonies (Obst, 1926; 1941). In that article, Obst dismissed the idea of European mass colonization of Africa and developed instead a plan for African over-production of raw materials for an autarkic Eurafrica (Heske, 1987b). All these concepts were based on a deep trust in the combination of geographical 'insights' and political power as well as a total ignorance of national rights to self-determination.

An important sub-field of German geopolitics became the so-called 'Wehrgeopolitik' (defence geopolitics), promulgated by Haushofer and other German geographers. This subfield of geopolitics stands in the tradition of German military science established by Clausewitz' book *Vom Kriege* (On War) (1832) and continued through the intimate links between geography and the military. Karl Haushofer and Oskar Ritter von Niedermayer, the exponents of 'Wehrgeopolitik' in the 1930s, were both high-ranking military officers before their academic careers in geography. Defence geopolitics, sometimes also called 'defence geography' or 'military geography' (Kost, 1988), was defined as the geographic study of 'the capability of a life form to assert itself'. An infamous highlight of this pseudo-science was the book *Raum und Volk im Weltkriege (Space and Nation in World War)* by the German geographer Ewald Banse (1932), which appeared in an English translation a year later under the title *Germany Prepares for War*. As a militarist analysis of physical and human landscape components, 'Wehrgeopolitik' contributed to a high degree to the German national mobilization for war.

Geopolitical thoughts, including colonial and war geography, became commonplace in German public schools during the Nazi period. Geopolitics became one of the basic elements of geographical education (Heske, 1988). 'Geopolitik' was to be a teaching principle penetrating geography and other school subjects such as history and biology. The National Socialist Teacher's League was founded in 1938, in addition to that of geography with its own 'Reichssachgebiet Geopolitik' (Imperial subject of geopolitics) to control, promote and educate a deeper geopolitical doctrine. The prominence of this kind of geopolitics stopped abruptly in June 1941, when Germany in 'Operation Barbarossa', attacked the Soviet Union. This invasion was the antithesis of the policy promoted in the global models of the German geopoliticians like Haushofer, who had pleaded for a pact with the Soviet Union in order to establish a powerful continental bloc in Eurasia (Haushofer, 1941). The decision of the Nazis to attack the Soviet Union demonstrated that the Nazis (and also the allies) used geopolitics only as a propaganda (and anti-propaganda) instrument and that it was never accepted as a policy instrument. Consequently the political importance of the pseudo-science, geopolitics, dropped dramatically in Germany from 1941 to 1945 and its main advocates were imprisoned or executed (Heske, 1987a).

Contra the geopolitics of war

The geopolitics of war could have developed only with the support of established academic geography. In the period 1870–1945, geopolitics encountered little resistance in the universities nor did it need to defend its position against any consistent criticism. The only fundamental criticism of geopolitical practice and discourse came from three 'outsiders', all geographers. Two were anarchists, Elisee Reclus (1830–1905) and Peter Kropotkin (1842–1921) and one a Marxist, Karl August Wittfogel (born 1896). All three were hounded from their home countries (France, Russia and Germany, respectively), Kropotkin and Wittfogel after imprisonment, and spending long periods in exile (Potter, 1984; Peet, 1985).

Reclus and Kropotkin met in 1877 in Switzerland and became lifelong friends. Kropotkin rejected realpolitics and developed an antithesis, his concept of an anarchist utopia (Breitbart, 1981). Reclus, though a sharp critic of colonialism, supported French population colonization in Algeria for a long time, viewing that country as an integral part of France (Giblin, 1987). Though he did not speak of 'geopolitics', he wrote in a geopolitical vein, in which he studied the interrelationships between physical and human phenomena (Lacoste, 1987). Reclus, for example, analysed the geographical organization of military systems which enabled them to control large colonial territories, going one step further and attacking these colonial methods of oppression, and he condemned exploitative colonization clearly and decisively. Reclus not only described economic and political exploitation of colonized countries but additionally questioned the consequences of this new development for the population as a whole, particularly denouncing policies which destroyed traditional structures with the backing of a privileged minority. In summary, the work of Reclus was somewhat of a contra-political geography, in opposition to the imperialist concept of Ratzel, but it was ignored by academic geography of the late nineteenth and early twentieth centuries.

The only forceful critic of German geopolitics was the sociologist, Karl August Wittfogel, who published his sharp attack on 'Geopolitik' in a Marxian broadsheet (Wittfogel, 1929). In that paper, Wittfogel demonstrated and undermined convincingly the imperialist implications of the writings of Richthofen, Ratzel, Kjellen and Haushofer. As a lone voice, he saw very clearly the political assistance of 'Geopolitik' to National Socialism, describing 'Geopolitik' in one passage as the attempt to provide a theoretical justification for the fascist tendencies of a matured imperialism (Wittfogel, 1929: 204–5).

World War II and the preceding decade were the heyday of geopolitics, not only in Germany and the USA, but also in Japan, Italy and other countries. Geopolitics served as a pseudo-scientific instrument for a war-promoting political propaganda. It tried to explain complex spatial-political connections through reductionist considerations. On the basis of a simplified explanatory model, geopoliticians provided prognoses about coming developments. The intellectual failure of geopolitics lies in its reductionism to a monocausal, often environmental, determinist approach. It is in the writings of modern American geostrategists that this monocausal model survives to the present.

American geopolitics

Certain premises underlay American 'democratic geopolitics' as it was described and justified by Bowman (1921) and Spykman (1942). First, the US was no longer content to remain an island power at the periphery of world affairs, relying on distance and an ocean location for protection and security. (See Sloan, 1988, for the change in geopolitical world-view, which Spykman was instrumental in promoting.) Second, the globe was viewed in a classic bi-polar mode, akin to the world of Mackinder's seapower/landpower juxtaposition, only now ideology was used to legitimate American activities on the perimeter of Eurasia, far from its home territories. Gaddis (1982) and Deibel and Gaddis (1987) detail the rhetoric used to substantiate the Truman Doctrine which was sold to a reluctant American public as an ideological, anti-Communist crusade. Third, as Sloan (1988) indicated, the theories of geopolitics, especially that of Spykman shaped the world-views of the policy-makers, leading them in turn to choose and promote a strategy that embodied the essence of the classical Heartland-Rimland juxtaposition. Though the evidence for this link is not over-whelming, there is little doubt that the immediate postwar policy analysts such as Kennan (1947) or later writers such as Gray (1988, 1989), Cline (1980), Luttwak (1983) and Brzezinski (1986) were strongly influenced by the writings of Spykman and frequently used the language of Mackinder (Heartland, pivot area, landpower versus seapower, etc). In a sense, just as the Nazis used the pseudo-theory of Haushofer's Geopolitik, so too, these strategists looked to classic theories for their justification of an already-formed global view.

There are five tenets of postwar American geopolitical strategy, all derived from Spykman's writings (Wilkinson, 1985; Gray, 1989). They are:

1 Anti-exceptionalism, arguing that the US should behave like all great powers viewing distant places and events as being integral to its interests and continued global position. (This was the antithesis of the pre-dominant nineteenth-century view of the US that saw the country as a noble experiment significantly different from the corrupt 'ancien regimes' of Europe, as noted by Agnew, 1983.) It is interesting that this viewpoint was accepted and promoted by the European emigrés who served as US National Security Advisers in the 1970s, Henry Kissinger and Zbigniew Brzezinski.
2 As a consequence of its global role, the US must necessarily be an interventionist state, contra isolationism, the dominant creed of the nineteenth century and that had its last hurrah after World War I when Congress refused to allow the US to join the League of Nations, despite President Wilson's admonitions.
3 Since the globe was becoming increasingly interdependent and the US was a global power, the US could therefore not be unaffected by developments anywhere and necessarily must be globalist (active on all continents but especially on the fringes of the world-island).
4 Adhering to a balance of power principle, the US could not allow any other power to achieve global hegemony and therefore set itself up as the anti-hegemonic state during World War II, in opposition to German and Japanese ambitions. After the war and in line with Mackinder's warnings about the domination of the Heartland by a single state, the US continued the anti-hegemonist principle despite the evidence that the US itself was clearly the global hegemon, controlling over 50 per cent of the world's GNP in the late 1940s (Williams, 1980). This theme is still alive and well (Wood, 1989). The US 'is the insular, ethical, reluctant actor in world affairs while the Soviet Union is the "village bully" ' (Gray, 1988, 54–5).
5 Spykman anticipated Containment, though he never used these words. Sloan (1988: 21–2), quoting Jean Gottmann, states that Spykman was

already warning the US strategic community in 1941, 3 weeks after Pearl Harbor, that the US would have to rely on the support of defeated Japan and Germany after the war in the alliance against the coming Heartland power, the Soviet Union. He basically accepted Mackinder's view that the greatest danger to the oceanic nations, including the US, was if the Heartland power managed to control the maritime nations of the Eurasian landmass. Therefore the US, in alliance with other Atlantic nations (linked by Mackinder's Midland = Atlantic Ocean), should preempt this possibility and control the Rimland through a string of alliances. Spykman and his postwar disciples envisioned the states of the Eurasian littoral as points that would be strung together along a line of alliances, like beads on a string (Sloan, 1988). Unlike Mackinder, who had a dynamic view of world politics and the landpower–seapower relationship, Spykman's (and his followers) view was essentially static. Despite the dramatic developments on the Eurasian littoral between 1945 and 1965 (especially the breakup of the European colonial empires, the anti-systemic Communist movements in many states and the Chinese–Soviet split in 1962), the Containment strategy remained unaltered. As summarized by Sloan (1988) and others such as Johnston (1985) and O'Sullivan (1986), strategy in the form of Containment was driving American foreign policy rather than the other way around, as had been originally envisioned in the late 1940s.

A central question about postwar American geostrategy is to what extent the policy and actions of the global hegemon were geographically invariate over time. Gaddis (1982), with his discussion of 'symmetry' and 'asymmetry' in American foreign policy, and Sloan (1988) both allude to this topic but neither tackles it directly. Sloan appears to believe that American actions were the result of 'indiscriminate globalism' after the Truman Doctrine of 1948 declared the whole globe as the arena of the Soviet–American competition. This phase changed to one of 'discriminate globalism' after the detente phase began in 1969 and Henry Kissinger became the most important foreign policy figure in Washington. Carter until 1979 also believed in an 'asymmetric' foreign policy until the Soviet invasion of Afghanistan and the Iranian revolution in 1979. From then until the first Reagan–Gorbachev summit in Geneva in 1985, we returned to the phase of 'indiscriminate globalism' in the Reagan Presidency, with its emphasis on reversing the pro-Soviet revolutions that were successful in the 1970s. An indicator of the Rimland strategy is the distribution of the incidents of Soviet and American 'gunboat diplomacy' since the war. Almost all the incidents were on the perimeter of the Soviet territory, a pattern that could suggest an expansionist American and defensive Soviet strategy (O'Loughlin, 1987).

The most definitive statement of a global strategy for the US in the future was recently published by Brzezinski, former National Security Adviser to

President Carter. Originally from Poland, he has always had an 'asymmetric' world-view since his early books on Soviet foreign policy. In sharp contrast to the static 'globalist' images of the Right, such as Cline (1980), Brzezinski's world is complex, regionally divergent, and demands a comparable American foreign policy. In his book *Game Plan: A Geostrategic Framework for the Conduct of the U.S.–Soviet Contest* (1986), Brzezinski uses the language of Mackinder and Spykman and offers a prescription for the US. His 'point of departure . . . is the geopolitical struggle for the domination of Eurasia' (p. xiii). He defines his geographic terms carefully so the uninformed in the ways of geopolitics can follow the argument. 'Geopolitics reflects the combination of geographic and political factors determining the condition of a state or region and emphasizing the impact of geography on politics' (p. xiv). This is, in fact, a narrower definition than Mackinder's and in our view precludes the incorporation and analysis of 'economic conditions'. This is geopolitics *sensu stricto* and it encompasses the seeds of its own problems. Similar views as represented in Parker (1988) and Gray (1988) cannot be accepted as adequate for the complex regionalist perspective demanded by an informed political geography.

Brzezinski, Gray (1988) and Luttwak (1983) believe power emanates from Eurasia: following Spykman, they believe that if the US remains isolated on the ocean periphery, without the use of a strong navy, global leadership will automatically default to the Soviet Union as the Heartland power. All three realistically accept that the US/Soviet contest is an imperial contest and are impatient with those who wish to give it the veneer of an ideological cloak. The Soviet–American rivalry is 'still the legatee of the old, almost traditional and certainly geopolitical clash between the great oceanic powers and the dominant land powers'. The US was in this sense the successor to Great Britain (and earlier Spain and Holland) and the Soviet Union to Nazi Germany (and earlier Imperial Germany or Napoleonic France). The seafaring states projected their power by exploiting the accessible ocean routes to establish transoceanic enclaves of political and economic influence. The land power sought continental domination as a point of departure for challenging the hegemony of the 'transoceanic intruder', (Brzezinski, 1986: 12). He dismisses sub-structural, including political economic, explanations in accounting for global success. 'It is geopolitical and strategic considerations that are critical in determining the focus, the substance, and eventually the outcome of this historical conflict' (Brzezinski, 1986: 11). In the geostrategists' view, containment was justified since the US had to become a continental (Rimland) power to prevent Soviet access to the Eurasian littoral. All accept Mackinder's famous dictum (quoted without attribution in Brzezinski's book) that whoever controls Eurasia controls world power. Gray (1988: 120) promotes a policy of 'dynamic containment' since the US/Soviet contest has become 'global' and that the old-fashioned ring around the Soviet Heartland is no longer sufficient.

Brzezinski's (1986) historical and geographic analysis of Soviet foreign policy conduct allows him to identify three Soviet prongs of expansion and as a result, demand a special American vigilance. They are termed the western (central Europe), far eastern (towards Korea and Japan) and south-western (towards Iran and the Persian Gulf) fronts. Like Kennan, Gray and Luttwak, Brzezinski sees the Soviet Union as historically expansionist (inheritors of the Great Russian tradition), defensively paranoid, and effectively landlocked, thereby demanding a projection to warm-water ports. Three sets of 'linchpin' states, defined as perimeter countries that 'are intrinsically important and in some sense up for grabs', are chosen and given special attention in their historical relations to the superpower competition. They are Germany and Poland in the west, South Korea and the Philippines in the East and either Iran or Afghanistan and Pakistan in the south-west. Buttressing their call to arms to the US and the Western community, Brzezinski (1986: 194) quotes Mackinder to the effect that 'democracy refuses to think strategically unless and until it has to do so for purposes of defence' while Gray (1988: 142) claims that 'the policy challenge for the United States is geostrategic, not ideological'. Since the US is placed most effectively where it (and its allies) are stronger (in Western Europe) and weakest where it is most vulnerable (in Southwest Asia), contemporary American geostrategists call for a global redistribution of US commitments and greater Allied sharing in terms of defence expenditures and regional assistance to protect the 'strategic interests' of the West.

The same kinds of criticisms that can be levelled at Brzezinski's strategic analyses can also be directed at political geographic works, such as those of Parker (1988). Space has no meaning except as a territorial container. All places are 'strategically important'; in effect, the globe is a strategic isotropic plain. These works attempt to be so geographic that they become ageographic. Cline (1980) advocates a global strategy for the US, though his model is based on oceanic control through domination of the globe's chokepoints and the development of an alliance of regional superpowers. There is little examination of the social, political and economic composition of areas nor consideration of any framework other than a very narrow power-political one.

Criticisms of geostrategic writings

The work of geostrategists can be criticized on two grounds, on strategy and on geographic methodology. We consider eleven shortcomings that we find are common in the writings of contemporary strategists.

1 The geostrategic view is ahistorical. The geostrategists accept that the Soviet–American competition is an imperial contest but do not recognize either how it became that way or the role of the US as the global hegemon since 1945. Perversely, they view the US as the anti-hegemonic state, trying to prevent the domination of Eurasia (and

consequently, the world) by a putative Soviet hegemon. This perversion of history has the aim of rationalizing the American presence on the Eurasian landmass but a long-cycle perspective, highlighting the geopolitical transition of global leadership from Britain to the US after World War II is much more persuasive. (See Taylor, 1988; Goldstein, 1988; and Modelski and Thompson 1988.) Klein (1988) demonstrates how the US has created and promoted a 'hegemonic culture' since World War II and the hegemon wearing anti-hegemonic disguises is a significant part of the image.

2 The geostrategists and other analysts of postwar American strategy (including the most comprehensive study by Gaddis, 1982) have seen the US as the reactor in response to Soviet initiatives. This thesis is unpersuasive and the historical evidence is replete with examples of American initiatives and later Soviet responses. A good example is the formation of NATO, which Kennan decried in 1948, followed later by the formation of the Warsaw Pact and the ossification of the bi-polar lines in Central Europe. Galtung (1989) claims that NATO is now 'autistic', operating a response system that ignores external reality and turns in on itself responding to its internal reactions. At a minimum, it is necessary to tease out the sequences of action–reaction by the two superpowers and an action–reaction model with both superpowers as initiators and respondents is much more persuasive (Rajmaira and Ward, 1990).

3 The geostrategists are correct when they describe the USSR as offering only a one-dimensional challenge (political-military) to the US. What they ignore are the other challenges to American global preeminence, coming in large measure from its security allies. In the era of detente in the summer of 1989, Americans told opinion pollsters that, by a ratio of 4–1, Japan constituted a greater threat to US security than the Soviet Union. With the stalemate in the US/Soviet military contest, it is most likely that the US will continue to spend 6–7 per cent of its GNP on the military, thereby reducing other necessary expenditures in capital forma-tion and infrastructure, thereby lending support to Kennedy's (1987) thesis of the causes of American relative decline. (See Oye *et al.* 1987; Strange, 1987; Goldstein, 1988; Lieber 1989; O'Loughlin, 1989; Waller-stein, 1987). By focusing exclusively on the Soviet Union as the only American challenger, the geostrategists have fallen prey to 'imperial overreach' and perversely act to exaggerate the American relative decline that they are anxious to prevent.

4 The perception of the Soviet Union by American geostrategists is consistent with the view that has dominated American foreign policy since 1945 and is enshrined in the agenda-setting NSC/68 document. Dalby (1988; 1990) analyses the concept of the Soviet 'Other' in contrast to the American perception of 'Self'. The 'Other' is dangerous, the evil empire, and the world is metaphysically structured as 'them' and 'us',

with us as superior. The images thus serve to ensure that militarization will continue and the legitimation of the Soviet–American contest is ensured (S. F. Cohen, 1985). It is worth remembering that President Truman could only persuade the American public to support his indiscriminate globalism through the device of portraying the USSR as the expansionist Communist power and its containment as an ideological necessity (Kolko and Kolko, 1972). President Gorbachev has turned the image of the Soviet Union on its head and has provoked a crisis of confidence among strategists. Some believe that the 'Gorbachev phenomenon' is just an abberration and so American vigilance, complete with high military expenditures must be maintained; others, including President Bush's advisors, have adopted a wait-and-see attitude; while a third group are willing to match Gorbachev's military reduction offers with a position of 'maximum flexibility' (Brzezinski, 1989). A smaller group has adopted the 'cornered bear' model, arguing that 'the more things fall apart in the Soviet Union, the more likely it is that Boris will make one last despairing lunge at the west' (Cockburn, 1989: 18).

5 Following on directly from the 'Other–Self' portrayal of the US/SU dialogue, the geostrategists see all states divided into two camps in a bipolar world. Nonalignment and neutrality have little role and are allowed only in places beyond the critical ring of competition around the Soviet Union. Consequently, a key element of the superpower competition is winning support of countries and denying control of strategically critical areas to the other side. In Brzezinski's and Gray's defence, it should be stated that they do not require the US to build an 'All-oceans alliance' (Cline, 1980) that demands that all states be classified as 'with us' or 'against us'. Nevertheless, for the linchpin areas, states are not allowed to choose a non-aligned path but are predetermined to serve a pawn-like role in the superpower confrontation.

6 Security for the geostrategists meets the classic power-political definition, comprised of control of territory, of building an unbeatable alliance of states in opposition to the Soviet bloc, of maintaining a lead in military hardware and troops, and supporting President Reagan's determination that the US will remain number one in the world. As noted by Ashley (1987), the 'security discourse' limits how individuals can act, write or speak about issues of war, peace and security. There are two dangerous fallacies with this narrow vision of security. First, in a zero-sum world view, security for the US is by definition insecurity for the USSR (Dalby, 1988) and thereby the arms and territorial race is heightened. Worse, because of the central location of the USSR in Eurasia, what is seen as a nodal position by the American strategists who follow the Mackinder model, is viewed as encirclement by the USSR (Parker, 1988: 140). Second, the narrow definition ignores the non-military definitions of security, including the provision of economic, health, educational and housing care for the citizens of the state. In a

country with hundreds of thousands homeless and the highest infant mortality rate in the Western world, this definition looks especially problematic. An even more worrying development is the broadening of the security net to incorporate protection of certain resources that are considered essential to the maintenance of the US standard of living or those of its allies. In this manner, the recent dispatch of US troops to the Persian Gulf in line with the Carter Doctrine (1980) was justified as protection of the oil-supplies to the West. It is generally agreed that the contest is unequal since the US, even by the Pentagon's admission, has a large lead in material base (ratio of total Warsaw Pact base to NATO base is about .30), equipment, technology and sophisticated weapons (Oneal, 1989; Herzfeld, 1989).

7 Brzezinski seems unaware that over time, greater restrictions have been placed on the use of US troops overseas both by congressional action and by unacceptance of such use by the American public, the so-called 'Vietnam syndrome'. It is generally agreed that no American president will sacrifice New York to save Hamburg in the event of a war outbreak in central Europe (Irving Kristol, quoted in Johnstone, 1984). Likewise, the legitimation of the use of American troops in overseas combat is becoming more difficult and consequently, restraints on the kinds of commitments that Presidents Truman, Kennedy and Johnson made are likely to prevent the dispatch of large-scale deployments to Brzezinski's 'linchpin zones'.

8 Turning specifically to the geographic methodology employed by American geostrategists, they are guilty of removing geography from their spatial perspective. Sayer's (1984: 52) comments on much contemporary social science research are particularly apt because strategists separate 'space and substance and speak of the effects of the uses of space . . . as if it were a thing existing independent of objects'. They talk about how superpowers can use territory (space) in their global rivalry: what they ignore is the content of what territorial units (peoples, materials, resources, histories, etc). Rosecrance (1987) shows how the superpowers are territorial states in a world of trading states: more territory and territorial control does not necessarily mean more power, as the geopolitical perspective demands. As a counter-example, the experience of Japan over the past two decades suffices. Lacoste (1984) has provided the most telling critique of this kind of geopolitics. Rejecting Napoleon's famous dictum that 'La politique d'un état est dans sa géographie', he argues that there is a hidden agenda behind the narrow geopolitical views. 'Recourse to "geographic" evidence and "implication" to justify the foreign policy of a state is a means of deflecting more complicated and less favourable analyses of that state's interests and ambitions' (p. 214). Lacoste is surely correct and deserving of support from political geographers when he states that 'geographic analysis must be multivariate, corresponding as it does, to the spatial configurations of economic,

demographic, social, political and cultural phenomena as well as various natural phenomena' (p. 218) and that the geographic 'conception of space is not that of the flat, uniform, and abstract space of the mathematician (and also often of the economist) but of a textual, extremely varied and very complex space' (p. 255). Brzezinski and other geopolitical writers have fallen victim to this perversion of the spatial perspective and the simplification of space.

9 Brzezinski follows the determinist tradition in geopolitics, starting with Ratzel and continuing with Haushofer and including contemporary writers like Gray and Cline. This group does not include Mackinder who was careful to consider the roles of policy-makers in the development of foreign policy. Mackinder also paid a great deal of attention to the historical relations between areas and the complex economic, social and cultural composition of areas that did not allow any simple determinist basis to foreign policy. 'The actual balance of political power at any given time is, of course, the product on the one level of geographical conditions, both economic and strategic, and on the other hand, of the relative number, virility, equipment and organization of the competing peoples' (Sir Halford J. Mackinder, 1904). States have no foreign policy: it is the state leaders that set it and they in turn respond to a whole set of factors, including both internal and external conditions. Recourse to the environmental model of international relations of the Sprouts (1965) allows us to incorporate both 'objective' conditions and 'subjective' (individual) factors. As Shapiro (1987) reminds us, the cartographical designation of states as geometric entities with precisely definable boundaries denies large parts of social reality. 'This discursive practice of reification is crucial to the operation of international relations theory and in particular to geopolitics' (Dalby, 1988: 417).

10 The geostrategists have to assume that the dyadic Soviet-American relationship will continue as at present, that is, in a nuclear stalemate but with a consistent ratcheting of arms developments and innovations. Under this nuclear hostage-taking relationship, they assume that no progress will be made in regional conflicts, in which the superpowers support different sides. While the first scenario is probably correct (little change in direct relations), recent developments cast doubts on the second (regional) scenario. Peace has broken out all over the world. Regional agreements have been signed in the past year in Angola/ Namibia, Afghanistan, Kampuchea, Iran/Iraq, Central America and the US is now talking to the PLO. Gorbachev is determined to cut Soviet losses in the Third World and has offered the US a regional quid pro quo, to pull out of Soviet bases in Vietnam in exchange for a US withdrawal from the Philippines. (For other developments in this area, see O'Loughlin, 1989.) In other words, all bets are off as Gorbachev turns Soviet expectations on their heads and the 'monolithic dinosaur' is

dismantled. Progress and agreements in bilateral relations may match similar progress in regional conflicts.

11 'Threat of peace cuts into arms merchants profits' ran a *Denver Post* headline on 18 December, 1988. The article went on, 'Western arms manufacturers, hit by a climate of peace in the world, are competing hard for markets in Asia, where they see a potential for strife.' This example indicates the 'warfare' state of many economies, including the US, with many localities dependent on employment in the arms industry (Kaldor 1988; Mintz and Ward 1989). Sanders (1983) terms US foreign policy as 'containment militarism', which has two legs, an interventionist policy abroad (see Johnston, 1985) and a militarized economy at home (Markusen, 1986). The 'empire as a way of life' doctrine (Williams, 1980) has a long history in American foreign policy, predating by a century the rise of the US to global leadership. The geostrategists do not question this legitimation but it is belatedly being recognized that there are alternative formulations to power status, including technological and economic leadership. (See the essays in Stoll and Ward, 1989.) When the major exports of the two leading military powers in the world include agricultural products, oil and minerals, while the leading trading states (Japan and West Germany, as well as the NICs) specialize in industrial and consumer goods exports, the substitution of military expenditure at the expense of economic investment obviously can be questioned. Boulding (this volume, Chapter 10), Kennedy (1987) and Thompson and Zuk (1986) warn about 'imperial overstretch' and its necessarily high military spending at the expense of other kinds of investments. The geostrategists, still locked in a cold war mindset, conventionally ignore the costs of the superpower contest in both present and future economic terms.

Conclusions

The term 'geopolitics', and especially its German counterpart 'Geopolitik', is strongly associated with power politics, imperialism, expanding territories and conflict. Therefore, some contemporary German geographers plead for avoidance of the term (Kost, 1986). In this chapter, we wish to show in which ways the geopolitical view could be undermined. With the exeption of Deudney (1983), there have been few attempts to develop a geopolitics of peace, though preliminary suggestions and research agendas are available (O'Loughlin and van der Wusten, 1986). A geopolitics of peace to us means a geographical and political science, which investigates global and regional social, political and economic processes in order to provide a foundation for conflict resolution and common security. It is essential that consideration of the operation of global-scale (structural) mechanisms as well as local conditions must be broadened to include power-political, ideological, geographic and economic explanations (Wallensteen, 1981).

As political geographers examine interstate relations, a useful start is to map and analyse dyadic relations on a cooperative-conflict scale (O'Loughlin and Anselin, 1990). Incompatibilities can be used to understand conflicts: common interests and securities can be used to understand cooperation. It is essential to extend consideration of conflicts beyond the military-political domain. As an example, consider the global map of Galtung (1985). The US is currently involved in four conflicts, with the Socialist bloc in political-military terms; with the Third World in an economic-political conflict of liberation from neo-colonial networks; with the fourth world in an economic competition which the fourth world has effectively won; and with its allies in the first world, resulting from the other three conflicts as the process of estrangement continues in the postwar Atlantic alliance. Galtung (1985: 60) concludes that 'no other country in the world is being projected to a process of status erosion that is so profound and so dangerous, so quickly'. And while this happens, the geostrategists fiddle to the Cold War tune of anti-Soviet confrontationism and territorial competition.

American security doctrines followed from classic geopolitical conceptions but without its language (Sloan, 1988). As the superpowers continue their efforts in the Third World, the American strategy is directed through (implicit and explicit) geographic analyses and policies such as the domino analogy, geostrategic control of continental blocks, regional challenges to US hegemony, ideological confrontation on world-order principles, respect of spheres of influence as codes of conduct and maintenance of established dyadic practices. As a result, security policy has not evolved substantially in the past 70 years. A focus on 'absolute space' has stripped regions of their characteristics, yielding an invariant foreign policy and an isotropic plain of strategy commitments. A few geographers working in the genre of strategic studies, like Saul Cohen (this volume, Chapter 2), have tried to show that regions matter but their efforts have fallen on the deaf ears of the so-called 'security community' (Vayrynen 1987, 1988).

Security has been defined traditionally as security from Others and defence of the state is viewed as the first responsibility of governments. Territorial control is the method of achieving security, both at home and abroad. In effect, it is a negative definition of security and if the concept of security is broadened to include non-military issues, the narrowness of the vision and the intellectual and practical limitations of geostrategic analyses can be easily identified. The importance of perception in determining positions and limiting resolution of political-geographic conflicts are illustrated in this volume by the chapters by Boulding and by Newman. We need to expand the definition of national security as the second part of the process of undermining geopolitics as it was and is practised. The intimate links between the domestic economy and the military needs of global powers are explored by Kaldor (1987) and she interprets the Second Cold War (1979–85) as an American attempt to reinstate the rigid bipolarity of the postwar years in a global economy of growing multipolarity and pluralism, which, from the outset, was doomed to failure (Wallerstein, 1989). Alternatively security conceptualizations, producing reductions in military expenditures, would result in a significantly altered domestic economy and possibly a renewed focus on basic needs. As suggested by the philosopher-physicist, Carl-Friedrich von Weizacker, what we need is not separate, often conflictual, foreign policies for each of the world's states but, instead, a domestic global policy for all the world's population.

Stable peace, the term proposed by Kenneth Boulding (1978), implies removal of all probable causes of conflict. Consequently, it is essential that basic needs be addressed as part of the definition of security. It is no geographic accident that the zones of stable peace in the world correspond to zones of wealth and surplus. It is doubtful that stable peace can be maintained in areas subjected to structural violence. Boulding's (1977) quarrel with Johan Galtung (1987) suggests that the term 'stable peace' be modified because there exists a mismatch between the concept (and its achievement) and Boulding's narrower vision of peace research as the study of military conflict and peace as non-war.

The study of international relations has not been able to inform us why some regions seem to have perpetual conflicts and others, with similar attributes, remain relatively peaceful. These issues are the stuff of the new political geography, which attempts to understand the global distribution of peace and security (broadly defined) (Starr and Siverson, 1990). Does the reduction of the friction of distance, allowing regional subsystems to form, promote more conflict or cooperation? What is especially needed is analysis and evaluation of the diffusion of conflict and cooperation, viewed best as two ends of the same dimension. Are certain kinds of neighbours more likely to be cooperative and are regions dominated by one state more peaceful than areas controlled by near equals. Are there plausible geographic solutions such as nuclear-weapons free zones? Regional disputes appear to remain intractable and it remains one of our tasks to understand the bases of the disputes and the social, political, environmental and economic legacies that they produce. Understanding, not national promotion, must be our aim as political geographers.

References

Agnew, J. A. (1983). An excess of 'national exceptionalism': towards a new political geography of American foreign policy. *Political Geography Quarterly,* **2**, 151–66.
Agnew, J. A. and S. Corbridge (1989). The new geopolitics: the dynamics of geopolitical disorder. In R. J. Johnston and P. J. Taylor (eds), *A World in Crisis: Geographical Perspectives.* 2nd ed., Oxford: Basil Blackwell, pp. 266–88.

Ashley, R. (1984). The poverty of neorealism. *International Organization,* **38**, 225–86.

Ashley, R. (1987). The geopolitics of geopolitical space: toward a critical social theory of international politics. *Alternatives,* **12**, 403–34.

Banse, E. (1932). *Raum und Volk im Weltkriege.* Oldenburg: Stalling.

Bassin, M. (1987). Imperialism and the nation state in Friedrich Ratzel's political geography. *Progress in Human Geography,* **11**, 473–95.

Boulding, K. (1977). Twelve friendly quarrels with Johan Galtung. *Journal of Peace Research,* **14**, 75–86.

Boulding, K. (1978). *Stable Peace.* Austin, Tx.: University of Texas Press.

Bowman, I. (1921). *The New World.* New York: World Book Company.

Bowman, I. (1942). The political geography of power. *Geographical Review,* **32**, 349–52.

Breitbart, M. (1981). Peter Kropotkin, the anarchist geographer. In D. R. Stoddart (ed.), Geography, Ideology and Social Concern. Oxford: Basil Blackwell, pp. 134–53.

Brzezinski, Z. (1986). *Game Plan: A Geostrategic Framework for the Conduct of the U.S.–Soviet Contest.* Boston: Atlantic Monthly Press.

Brzezinski, Z. (1989). *The Grand Failure: The Birth and Death of Communism in the Twentieth Century.* New York: Charles Scribner's Sons.

Cline, R. C. (1980). *World Power Trends and U.S. Foreign Policy in the 1980s.* Boulder, CO.: Westview Press.

Cockburn, A. (1989). Ron's parting shots. *New Statesman,* 13 January, pp. 18–19.

Cohen, S. B. (1982). A new map of global geopolitical equilibrium: a developmental approach. *Political Geography Quarterly,* **1**, 223–42.

Cohen, S. F. (1985). *Rethinking the Soviet Experience: Politics and History since 1917.* New York: Oxford University Press.

Dalby, S. (1988). Geopolitical discourse: the Soviet Union as 'Other'. *Alternatives,* **13**, 415–42.

Dalby, S. (1990). American security discourse: the persistence of geopolitics. *Political Geography Quarterly,* **9** (forthcoming).

Deibel, T. L. and J. L. Gaddis (eds) (1987). *Containing the Soviet Union: A Critique of US Policy.* Washington, DC: Pergamon-Brassey's International Defense Publishers.

Deudney, D. (1983). *Whole Earth Security: A Geopolitics of Peace.* Washington DC: World Watch Institute.

Faber, K. G. (1982). Zur Vorgeschichte der Geopolitik: Staat, Nation und Lebensraum im Denken deutscher Geographen vor 1914. In *Weltpolitik, Europagedanke, Regionalismus.* Munster: Aschendorff, pp. 389–406.

Gaddis, J. L. (1982). *Strategies of Containment: A Critical Appraisal of Postwar American National Security Policy.* New York: Oxford University Press.

Galtung, J. (1985). Military formations and social formations: a structural analysis. In P. Wallensteen, J. Galtung and C. Portales (eds), *Global Militarization.* Boulder, CO.: Westview Press, pp. 23–74.

Galtung, J. (1987). Only one friendly quarrel with Kenneth Boulding. *Journal of Peace Research,* **24**, 199–203.

Galtung, J. (1989). The Cold War as an exercise in autism: the US government, the governments of Western Europe and the people. *Alternatives,* **14**, 169–93.

Giblin, B. (1987). Elisee Reclus and colonisation. In P. Girot and E. Kofman (eds.), *International Geopolitical Analysis*. London: Croom Helm, pp. 26–45.

Goldstein, J. (1988). *Long Cycles*. New Haven, CT: Yale University Press.

Gray, C. S. (1988). *The Geopolitics of Super Power*. Lexington, KY: University Press of Kentucky.

Gray, C. S. (1989). Ocean and continent in global strategy. *Comparative Strategy* 7, 439–44.

Hansch, B. F. (1912). Die Aufteilung Afrikas. *Geographische Zeitschrift*, **18**, 361–87.

Haushofer, K. (1941). *Der Kontinentalblock: Mitteleuropa, Eurasien, Japan*. Munich: Eher.

Herodote (1987). Geographers, action and politics. In P. Girot and E. Kofman (eds.) *International Geopolitical Analysis*. London: Croom Helm, pp. 1–9.

Herzfeld, C. (1989). Technology and national security: restoring the U.S. edge. *Washington Quarterly*, **12**, 171–83.

Heske, H. (1987a). Karl Haushofer: his role in German geopolitics and in Nazi politics. *Political Geography Quarterly*, **6**, 135–44.

Heske, H. (1987b). Der Traum von Afrika: zur politischen Wissenschaftsgeschichte der Kolonialgeographie. In H. Heske (ed.), *Ernte-Dank?: Landwirtschaft zwischen Agrobusiness, Gentechnik und traditionellem Landbau*. Giessen: Focus, pp. 204–22.

Heske, H. (1988). '. . . und morgen die ganze Welt'. *Erdkundeunterricht im Nationalsozialismus*. Giessen: Focus.

Heske, H. and R. Wesche (1988). Karl Haushofer 1869–1946. In T. W. Freeman (ed.) *Geographers: Bibliographical Studies, 12*. London: Mansell (forthcoming).

Holsti, O. R. and J. N. Rosenau (1986). Consensus lost. Consensus regained? Foreign policy beliefs of American leaders, 1976–1980. *International Studies Quarterly*, **30**, 375–409.

Hudson, B. (1977). The new geography and the new imperialism, 1870–1918. *Antipode*, **9**, 12–19.

Johnston, R. H. (1985). Exaggerating America's stakes in Third World conflicts. *International Security*, **10**, 32–68.

Johnstone, D. (1984). *The Politics of Euromissiles: Europe's Place in America's World*. London: Verso.

Kaldor, M. (1987). The world economy and militarization. In S. Mendelovitz and R. B. J. Walker (eds), *Towards Just World Peace: Perspectives from Social Movements*. London: Butterworths, pp. 49–78.

Kaldor, M. (1988). Transforming the state: an alternative security concept for Europe. In B. Hettne (ed.), *Europe: Dimensions of Peace*. London: Zed Books, pp. 204–20.

Kennan, G. (Mr. X) (1947), The sources of Soviet conduct. *Foreign Affairs*, **25**, 566–82.

Kennedy, P. M. (1987). *The Rise and Fall of the Great Powers: Economic Change and Military Conflict from 1500 to 2000*. New York: Random House.

Klein, B. S. (1988). Hegemony and strategic culture: American power projection and alliance defense politics. *Review of International Studies*, **14**, 133–48.

Kolko, J. and Kolko, G. (1972). *The Limits of Power: The World and United States Foreign Policy*. New York: Harper and Row.

Kost, K. (1986). Begriffe und Macht: Die Funktion der Geopolitik als Ideologie. *Geographische Zeitschrift*, **74**, 14–30.

Kost, K. (1988). *Die Einflusse der Geopolitik auf Forschung und Theorie der Politischen Geographie von ihren Anfangen bis 1945*. Bonn: Bonner Geographische Abhandlungen no. 76.

Krings, T. (1984). Einhundert Jahre Berliner Kongo-Kongferenz von 1884–1885: ein Ruckblick auf die Hintergrunde, Ergebnisse und Folgen. *Die Erde*, **115**, 295–304.

Lacoste, Y. (1976). *La géographie, ca sert, d'abord, à faire la guerre*. Paris: Maspero.

Lacoste, Y. (1984). Geography and foreign policy. *SAIS Review Ser 2*, **4**, 213–27.

Lacoste, Y. (1987). The geographical and the geopolitical. In P. Girot and E. Kofman (eds), *International Geopolitical Analysis*. London: Croom Helm, pp. 10–25.

Lieber, R. J. (1989). Eagle revisited: a reconsideration of the Reagan era in U.S. foreign policy. *Washington Quarterly*, **12**, 115–26.

Luttwak, E. N. (1983). *The Grand Strategy of the Soviet Union*. London: Weidenfeld and Nicholson.

Mackinder, H. J. (1904). The geographical pivot of history. *Geographical Journal*, **23**, 421–37.

Markusen, A. (1986). Defense spending as industrial policy. *International Journal of Urban and Regional Research*, **10**, 405–22.

McKay, D. V. (1943). Colonialism in the French geographical movement 1871–81. *Geographical Review*, **33**, 214–32.

Mintz, A. and M. D. Ward (1989). The political economy of military spending in Israel. *American Political Science Review*, **83**, 521–33.

Modelski, G. and W. R. Thompson (1988). *Seapower in Global Politics, 1494–1993*. Seattle: University of Washington Press.

Obst, E. (1926). Wir fordern unsere Kolonien zuruck! *Zeitschrift fur Geopolitik* **3**, 152–60.

Obst, E. (1941). Ostbewegung und afrikanische Kolonisation als Teilaufgaben einer abendlandischen Grossraumpolitik. *Zeitschrift fur Erdkunde*, **9**, 265–78.

O'Loughlin, J. (1987). Superpower competition and the militarization of the Third World. *Journal of Geography*, **86**, 269–75.

O'Loughlin, J. (1988). Is there a geography of international conflicts. *Political Geography Quarterly*, **7**, 85–91.

O'Loughlin, J. (1989). World-power competition and local conflicts in the Third World. In R. J. Johnston and P. J. Taylor (eds), *A World in Crisis: Geographical Perspectives*. 2nd edn, Oxford: Basil Blackwell, pp. 289–332.

O'Loughlin, J. and L. Anselin (1990). Bringing geography back to the study of international relations: spatial dependence and regional context in Africa, 1966–1978. *International Interactions*, **16** (forthcoming).

O'Loughlin, J. and H. van der Wusten (1986). Geography, war and peace: notes for a contribution to a revived political geography. *Progress in Human Geography* **10**, 484–510.

O'Loughlin, J. and H. van der Wusten (1990). The political geography of Panregions: the theory and an empirical example of Eurafrica. *Geographical Review*, **80**, (forthcoming).

Oneal, J. R. (1989). Measuring the material base of the contemporary East-West balance of power. *International Interactions*, **15**, 177–96.

O'Sullivan, P. (1986). *Geopolitics*. London: St Martin's Press.

Oye, K., Lieber, R. J. and Rothchild, D. (eds) (1987). *Eagle Resurgent: The Reagan Era in American Foreign Policy*. Boston: Little Brown.

Parker, G. (1988). *The Geopolitics of Domination.* London: Routledge.

Peet, R. (1985). Introduction to the life and thought of Karl Wittfogel, with an appendix on the Asiatic mode of production. *Antipode,* **17**, 3–20.

Potter, S. R. (1984). Peter Alexeivich Kropotkin 1842–1921. In T. W. Freeman (ed.) *Geographers, Bibliographical Studies,* **7**. London: Mansell, pp. 63–9.

Raimaira, S. and Ward, M. D. (1990). Evolving foreign norms: reciprocity in the superpower triad. *International Studies* Quarterly, **34** (forthcoming).

Ratzel, F. (1896). Die Gesetze des raumlichen Wachstums der Staaten. *Petermanns Mitteilungen* **42**, 92–107. Reprinted in J. Matznetter (ed.), *Politische Geographie.* Darmstadt: Wissenschaftliche Buchgesellschaft, 29–53.

Ratzel, F. (1897). *Politische Geographie oder die Geographie der Staaten, des Verkehrs und des Krieges.* Munich: Oldenbourg.

Rosecrance, R. (1987). Long cycle theory and international relations. *International Organization,* **41**, 283–301.

Sanders, J. (1983). *Peddlars of Crisis: The Committee on the Present Danger and the Politics of Containment.* Boston: South End.

Sandner, G. (1989). The 'Germania triumphans'-syndrome and Passarge's 'Erdkindliche Weltanschauung': roots and effects of German political geography beyond 'Geopolitik'. *Political Geography Quarterly,* **8** (forthcoming).

Sayer, A. (1984). *Method in Social Science: A Realist Approach.* London: Hutchinson.

Schultz, H. D. (1987). Pax Geographica.: Raumliche Konzepte fur Krieg und Frieden in der geographischen Tradition. *Geographische Zeitschrift,* **75**, 1–22.

Shapiro, M. (1987). *The Politics of Representation.* Madison: University of Wisconsin Press.

Sloan, G. R. (1988). *Geopolitics in United States Strategic Policy, 1890–1987.* Brighton: Wheatsheaf Books.

Smith, N. (1984). Isaiah Bowman: political geography and geopolitics. *Political Geography Quarterly,* **3**, 69–76.

Smith, N. (1986). Bowman's New World and the Council on Foreign Relations. *Geographical Review,* **76**, 438–60.

Smith, W. D. (1986). *The Ideological Origins of Nazi Imperialism.* New York: Oxford University Press.

Sprout, H. and Sprout, M. (1965). *The Ecological Perspective on Human Affairs, with Special Reference to International Politics.* Princeton, NJ: Princeton University Press.

Spykman, N. J. (1938). Geography and foreign policy I. *American Political Science Review,* **32**, 28–50.

Spykman, N. J. (1942). *America's Strategy in World Politics.* New York: Harcourt, Brace & Company.

Stoll, R. J. and Ward, M. D. (eds) (1989). *Power in World Politics.* Boulder, CO.: Lynne Rienner.

Strange, S. (1987). The persistent myth of lost hegemony. *International Organization,* **41**, 551–74.

Starr, H. and Siverson, R. M. (1990). Alliances and geopolitics. *Political Geography Quarterly,* **9** (forthcoming).

Taylor, P. J. (1985). The value of a geographical perspective. In R. J. Johnston (ed.) *The Future of Geography.* London: Methuen, pp. 92–110.

Taylor, P. J. (1988). *Geopolitical Transition*. Paper presented at the Annual Meeting of the Association of American Geographers, Phoenix, Arizona, April.

Thompson, W. R. and Zuk, G. (1986). World power and the strategic trap of territorial commitments. *International Studies Quarterly, 30*, 249–67.

Toal, G. (1989). *Geopolitics as the Writing of Space and Place*. Unpublished Ph.D. dissertation, Department of Geography, Syracuse University.

Vayrynen, R. (1984). Regional conflict formations: an intractable problem of international relations. Journal of Peace Research, 21, 337–59.

Vayrynen, R. (1988). East-West rivalry and regional conflicts in the Third World. In B. Hettne (ed.), *Europe: Dimensions of Peace*. London: Zed Books, pp. 101–26.

Wallensteen, P. (1981). Incompatibility, confrontation and war: four models and three historical systems. *Journal of Peace Research, 18*, 57–90.

Wallenstein, I. (1987). The Reagan non-revolution, or the limited choices of the U.S. *Millenium, 16*, 467–72.

Wallerstein, I. (1989). The capitalist world-economy: middle term prospects. *Alternatives, 14*, 279–88.

Wilkinson, D. (1985). Spykman and Geopolitics. In C. Zoppo and C. Zorgbibe (eds), *On Geopolitics: Classical and Nuclear*. Dordrecht: Martinus Nijhoff, pp. 77–130.

Williams, W. A. (1980). *Empire as a Way of Life*. New York: Oxford University Press.

Wittfogel, K. A. (1929). Geopolitik, geographischer Materialismus und Marxismus. *Unter dem Banner des Marxismus, 3*, 17–51. Cited after the reprint in J. Matznetter (ed.) *Politische Geographie*. Darmstadt: Wissenschaftliche Buchgesellschaft, pp. 183–232.

Wood, R. S. (1989). Strategic choices, geopolitics and resource constraints. *Washington Quarterly, 12*, 139–56.

Zoppo, C. and Zorgbibe, C. (eds) (1985). *On Geopolitics: Classical and Nuclear*. Dordrecht: Martinus Nijhoff.

4 The Geopolitics of Dominance and International Cooperation

Geoffrey Parker

That particular geographical surface which Henrikson called 'the neatly segmented multicolored world of the standard political map' overlies and, from a geopolitical perspective, obscures, the physical, economic, social and cultural surfaces. Like magma, this surface shows a marked tendency to solidify into a hard crust above the others. This is potentially its most dangerous characteristic, since it impedes that more gradual adjustment to change which has characterized the world's other anthropogeographic surfaces (Henrikson, 1980). As a result pressures from beneath have frequently built up, resulting in periods of explosive violence. It is this feature of the geopolitical surface which has encouraged attempts to replace the system of sovereign territorial states with a more satisfactory alternative. Historically the two most favoured alternatives have been either dominance or cooperation. Dominance entails the replacement of the polychrome world by a monochrome one, while cooperation on the other hand entails the surrender or merging of aspects of state sovereignty for the common good. Each of these options has aspired ultimately towards some form of universality, but initially each has arisen within a specific context.

What Martin Luther termed the *animus dominandi*, the lust to control others, is seen by Patrick O'Sullivan as being 'the driving force of history' (O'Sullivan, 1986). A state imbued with this lust eventually attempts to replace the existing order with a new order centring upon itself. The existing political units will come to be treated as impediments to progress – Kleinstaatengerumpel – ripe either to be incorporated into the new 'greater' state or to be reduced to the status of satellites. A number of such dominant systems have been created in the western ecumene during modern times, but not one of them has been completely successful in replacing the existing structure with a lasting alternative. A residue of the sovereign state structure has always remained in being; it has often consisted of no more than the regional periphery, but this periphery has been able to summon up combined strength sufficient to challenge the dominant state and eventually to bring about its downfall. In its bid to achieve dominance the expanding state embraces an ideology which both gives direction to its expansionist drive and at the same time also seeks to legitimize the expansionist process. The ideology further serves to underpin the new order when this has been, at least partially, established. As Patrick O'Sullivan puts it, this is the 'cloak of righteousness' which clothes 'the lust for power', and it is used to justify the acquisition of Lebensraum, the fulfilment of 'Manifest Destiny' or the satisfaction of the imperatives of historical dialectic (O'Sullivan, 1986).

Dominant states possess a number of clearly identifiable spatial characteristics which appear to be inherent in the expansionist process itself. Important among them are the geopolitical circumstances of the origins of the state, the evolution of its core region, the development of its core–periphery structure, the role of the state in nation development, and its relationship to the overall culture of which it forms a part (Parker, 1987). Together such characteristics may be taken to constitute the diachronic spatial structure of the dominant state, and from the characteristics of those states which have made bids for dominance in the Western ecumene in modern times I have constructed a spatial model and proposed a geopolitical theory of the process of dominance. The degree of concurrence with the model has proved to be a very close one. Bids for dominance have normally taken place at times when the geopolitical surface is at its most rigid, and when the possibility of gradual change is consequently most remote.

The other method of attempting to produce a more stable geopolitical surface has been the establishment of organizations intended to regulate interstate, or international, relations. Historically, attempts to establish such organizations have been most in evidence in the Western ecumene in the immediate aftermath of general wars, that is to say after a particular bid for dominance has been defeated. While some success has been achieved in the ordered regulation of interstate relations, its effectiveness has always been impeded by the unwillingness of the great powers to give up what they consider to be their rights to unilateral action in pursuit of their own basic self-interest (Parker, 1987). While small states have usually been more likely to see themselves as being gainers from an effective international system, the great powers have always tended to behave more like gamblers, keeping their options open and cherishing the hope of being ultimate gainers from chance and uncertainty on the international scene. The link between dominance and uncontrolled sovereignty is that it is from the ranks of the existing great powers that the next bidder for the position of dominance has most often arisen. On the whole question of bringing about greater stability to the world's geopolitical surface the great powers have thus by their behaviour tended to favour the option of dominance rather than that of cooperation.

Since the spatial structures of the dominant states of the past have accorded closely with a geopolitical model of dominance, it should thus be possible to identify present or future contenders for dominance by an examination of contemporary spatial structures. This entails a prediction of likely future spatial patterns from the evidence provided by spatial patterns of the past. The modification or alteration of such patterns then becomes essential if the recurring cycles of domination, and the conflicts which result from them, are to be brought to an end. Past response to the general perception of the rise of a proto-dominant state has been the attempt to assemble an alliance capable of holding its ambitions in check. However, even if this is successful, and historically in the Western ecumene this has been the case, it has a fundamental flaw as a method for bringing stability to the system. This is that it is an essentially ad hoc reaction to events which are already taking place; the disease has become established in the body politic before the correct diagnosis has been made and the necessary treatment prescribed. War rather than peace is by then the only method available for defending the existing geopolitical order.

The identification of a new proto-dominant state at an early stage in its evolution through the examination of its diachronic spatial structure and its proximation to the geopolitical model of dominance would certainly make it easier for early countervailing action to be taken. However, to be effective such action has to entail something more positive than the scraping together of ad hoc alliances. It necessitates the development of an alternative strategy to the dominant state tendency. International organizations of the traditional type have a built-in predisposition to revert to the dominant state model. The strategy has to be based on an organization which is able to oblige its members to abide by decisions collectively reached. Clearly organizations which possess this power must be suprastate rather than interstate in their structures. Since World War II a number of organizations of regional cooperation have been established, and the most far-reaching of these have been in Europe. The most wide-ranging and evolutionary in character is the European Community and many areas of the relationships of its member countries have become subject to joint decision-making. Its initial objective was the achievement of security and its ultimate objective the holistic one of complete integration.

The community method was constructed on the three pillars of regionalism, supranationalism and functionalism. In the particular form adopted it arose from Jean Monnet's concept of gradual functional accretion, of the achievement of 'peace by pieces' (Parker, 1983). Specific areas of activity having important security implications were brought under the control of a body which was given the powers necessary to regulate and harmonize their functions. Following the initial successes, new and related areas were brought into the sphere of competence. The actual areas were largely a matter of trial and error, of building pragmatically on actual manifest achievement. New members then joined the original community countries. The method of 'peace by pieces' thus entailed two processes inextricably linked together. One was functional, the other spatial. The functional process endeavoured to fuse the geopolitical surface to the economic, social and other surfaces, thus treating them as being essentially interdependent. The spatial process entailed the gradual extension of the Community over the geopolitical surface itself. From a specifically geopolitical perspective the truly innovative character of the Community arises both from this recognition of the necessity to treat the various surfaces holistically, and from the consideration of this as being part of a more radical process leading eventually to complete integration. At its most absolute it is thus firmly post nation-state in its thinking, but the step-by-step approach inherent in the achievement of 'peace by pieces' allows the bastions of the state to be taken by steady sapping rather than by dramatic and probably ineffectual, frontal assault.

The success of the community method in a small part of Europe in the aftermath of World War II was favoured by a particular set of geopolitical circumstances. Fundamental was the manifest failure of the existing state system of the region either to provide for the overall wellbeing of its peoples or to prevent the repeated emergence of dominant states within it. As a result, both the ideology of nationalism and the dominant state ideology were equally discredited, and particular opprobrium was attached to that nation state which had most recently sought to overthrow the system and to replace it with its own particular brand of dominance. The disenchantment was made more absolute by the imminent disintegration of the Eurocentric world system and the shift of global power away from north-west Europe for the first time in three centuries. The original European Community did constitute in a very real sense what William Pickles labelled 'a league of defeated countries' and it was this sense of 'defeat' on a broad front which facilitated the paradigm shift from confrontation to cooperation. The rough balance which existed among its erstwhile great powers, now very much cut down to size, created a 'window of opportunity' for cooperation and proved for a time conducive to the establishment of a suprastate organization. The draining of power in the international arena was accompanied by a pronounced tendency for the devolution of power within the state from the centre to the periphery. The geopsychology of the latter was more akin to that of small states than to that of the still often imperially minded core regions. The six founder members of the Community served as that 'core of strength' which Karl Deutsch saw as being the prerequisite and centre of gravity for any successful regional organization (Merritt, 1974). At the very heart of this 'core of strength' there was a transnational proto-core possessing many historical, cultural and economic features of a transnational character. Formerly the centre of regional rivalry and conflict, the greater Rhineland was to become the focus of the forces which aspired towards unity. Whatever the bonuses arising from cooperation, its greatest achievement was the construction of what Galtung called a 'peace structure' in which the states could live together free from the threat of dominance. 'It is peace built by associative means', said Galtung, 'and certainly on a stronger foundation than attempts earlier this century to base peace on dissociation, balance of power, on Maginot and Siegfried Lines' (Galtung, 1981). The relationship of the whole process to the world geopolitical surface was also of considerable significance. This focused upon the dichotomy between the two new centres of world power. From a Community perspective one of these was perceived as being hostile and a possible bidder for a position of dominance, while the other was perceived to be largely benign and was encouraged to assume the role of guarantor of the process and of the security of those states engaged in it from threats of dominance whether arising from within or from without. These were the essential geopolitical conditions which underlay the willingness of the core countries of the grouping to envisage a future in which association rather than dissociation would be the norm, and to embark upon the creation of structures appropriate to this objective. Since the time of its creation a great power psychology has once more emerged in places, but the momentum which began at the time of the 'window of opportunity' has proved sufficiently strong to withstand such attitudes.

The successful replication of this community method, the achievement of 'peace by pieces' by means of closer regional interdependence, in other parts of the world must depend upon the existence of a sufficiency of those geopolitical conditions which underlay its success in Europe. In attempting to identify the existence of such conditions one would look for a group of contiguous states having the potential for increased cooperation, so becoming a 'core of strength'. Such a group would ideally constitute an 'island of relative similarity', as Russett put it, in a diverse and 'variegated' world. Such 'similarity' is not considered by Russett to be an absolute precondition, but if it is absent then co-operation can only be attained through 'heavy communication and interaction', (Russett, 1967). Another desirable condition is the presence of a regional proto-core around which a 'third identity' can begin to emerge. Cuneo sees such an area as having been historically one of conflict, but emerging as the 'tragic birthplace' of the new unit which in time grows to encompass the existing states (Cuneo, 1963). Further positive conditions include the existence of a multiplicity of centres of power which are able to balance one another, thus discouraging the rise of one of them to a position of dominance; there should also be the possibility of some intra-state restructuring so as to bring about a further devolution of power within broader cultural formations. The integrity of the process could best be maintained by the presence of an external great power which could act for a time as its guarantor. Such conditions form a plausible basis firmly founded in geopolitical realities for a shift from dissociative to associative means for the achievement of lasting peace and security. These conditions must exist in a state of favourable concurrence, producing a 'window of opportunity' which can be used to initiate the process. Haas has pointed out that an essential prerequisite is the existence of the political will for change, and this entails above all the conviction of the respective elites that it is going to be in their own broader interests (Haas, 1964). If this conviction is not present, then the integrative method of cooperation stands little chance of success even if the other conditions are favorable.

However, future world geopolitical scenarios based upon past experience in the Western ecumene must realistically still include the possibility of the rise of new dominant states from within the ranks of the regional great powers. Since World War II, hegemonial powers have been replacing the former empires in various parts of the world and it is the replication of this geopolitical characteristic of the Western ecumene, rather than the relatively recent tendency to cooperation, which appears to be all to prevalent. Yet the rise of hegemonial power is clearly not compatible with the development of effective strategies of security and cooperation. Great power has been always achieved and consolidated by force and the threat of force rather than by its abrogation. To be effective any regional strategy must entail the participation of all states within the region. More than this, if it is to be an effective alternative to the states, it must entail the integration of state power into the wider regional entity, thus leading to the emergence of Cuneo's 'third identity'. Not only are dominance and cooperation clearly incompatible but, as has been seen, each has its own optimum geopolitical structure. Quite clearly such structures are mutually exclusive in any given territory. The creation of the geopolitics of cooperation can only be achieved by the disbandment of the geopolitics of domination. The success of the one inevitably means the failure of the other.

Historically, the geopolitics of domination has constituted the norm in the Western ecumene. There have been all two few examples of the geopolitics of cooperation. However, in recent times, change has been taking place which makes it more realistic to envisage a new geopolitics based upon co-operative principles. Halford Mackinder, seventy years ago, saw all too few signs of any real convergence of idealism and reality in the international arena. On the basis of experience since *Democratic Ideas and Reality* was written such convergence now appears to be more of a possibility. To be successful in the long term this entails nothing less than the complete reconstitution of the world's geopolitical surface and the recombination of its segments into larger and more effective groupings. It also entails introducing fluidity into a surface which has had a history of rigidity. Herein lies the importance of being able to identify contemporary proto-dominance and to act accordingly. In the past cooperation has been but a weak and belated reaction to dominance. A cooperative strategy based upon an effective alternative geopolitical model encapsulates also an alternative spatial ideology. This must give hope to those opposed to dominance not only in the small nation states which have so much to lose but also to those within the great powers themselves who have steadfastly opposed the ideology of dominance but have not been successful in pointing their fellows towards alternatives to it.

'We must pray', wrote J. M. Keynes in 1919, 'that the souls of the European peoples turn away this winter from the false idols . . . and substitute in their hearts for the hatred and the nationalism, which now possess them, thoughts and hopes of the happiness and solidarity of the European family' (Commission of the European Communities, 1967). On 22 May 1937 – nearly two decades later – the long-running Disarmament Conference, which had by that time come down to arguing about the number of bolts and nuts in a tank, regretfully announced its indefinite adjournment. On the following day the David Low cartoon in the London *Evening News* showed the great powers as tigers, lions and jackals announcing to the assembled peoples of the world, represented by a flock of sheep, through their spokesman, a crocodile, 'My friends, we have failed. We just couldn't control your warlike passions' (Low, 1937). The geopolitics of domination had once more triumphed in the world.

All ideals, as Mackinder wished his airy cherub to remind the statesmen assembled at Versailles, have to be founded on reality if they are to stand any chance whatsoever of being successful (Mackinder, 1919). 'Thoughts and hopes' and pleas to 'turn away . . . from false idols' have never proved to be sufficient. A definitive shift away from the geopolitics of domination towards the geopolitics of cooperation now constitutes both a great challenge and a great opportunity for the future. If this is to be successfully accomplished, then geographical reality must not only be kept always clearly in mind, but must remain a firm guide both to action and to the ultimate policy objectives.

References

Commission of the European Communities, *European Community* London, July 1967.
Cuneo, J. (1963). *Science and History*. London: Cassell.
Galtung, J. (1981). *The European Community: A Superpower in the Making*. London: Allen and Unwin.
Haas, E. B. (1964). The Balance of Power. In W. A. Douglas Jackson, *Politics and Geographic Relationships*. Englewood Cliffs: Prentice Hall, pp. 465–84.
Henrikson, A. K. (1980). The Geographical Mental Maps of American Foreign Policy Makers, *International Political Science Review: Politics and Geography*. I, 495–530.
Low, D. (1937). The Conference Excuses Itself, Cartoon, *Evening Standard*, London, 23 May, 1937.
Mackinder, H. J. (1919). *Democratic Ideals and Reality*. London, Constable.
Merritt, R. (1974). Locational aspects of political integration. In K. R. Cox, D. R. Reynolds and S. Rokkan (eds), *Locational Approaches to Power and Conflict*. New York: Wiley, pp. 187–212.
O'Sullivan, P. (1986). *Geopolitics*. London: Croom Helm.
Parker, G. (1983). *A Political Geography of Community Europe*. London, Butterworths.
Parker, G. (1987). *The Geopolitics of Domination*. London: Routledge.
Russett, B. (1967). *International Regions and the International System: A Study in Political Ecology*. Chicago: Rand McNally.

5 The Growing Problem of Incomplete Surrenders; 'Neither War nor Peace and Its Geographic Implications'

George H. Quester

One of the more subtle and difficult concepts for an analyst of international politics to handle may be that of 'surrender'. What have the two sides done, with regard to the future of international relations when one of them surrenders to another as a means of terminating an armed conflict? Our normal intuition is that the loser must grieve and the victor must rejoice. But there have been winners who were disappointed later with what such a transaction had brought them, and there have been 'losers' who began immediately scheming to come back to revise the terms of the peace (there have also been losers who were disappointed by how little opportunity was to be retained for any such comeback and there have been victories that were so total that they become enshrined as the new status quo of history).

As illustrated across the cases which follow, both the victors and the vanquished, at the moment of a surrender, must be burdened with some lingering uncertainties as to what is now slated to happen. As Trotsky and his fellow Bolsheviks attempted to make a meaningless surrender to the Germans in 1917, in a policy of 'neither peace nor war' (Wheeler-Bennett, 1939) they may have offered a model to the Arab states fifty years later, whenever such states were confronted with superior Israeli military power (Harkabi, 1972). And they may also offer a model of how established states across the globe will have to deal with terrorist hostage-takings and other hostage threats in the future. We will thus have to reconsider the distinctions sometimes preferred between a 'just peace' and a 'victor's peace', and between a genuine surrender and a feigned surrender. We will have to become discontent here with what our intuitions more naively would have told us was the notion of a surrender, re-examining the very essence of what the contract of a capitulation actually must amount to.

Some examples for analysis

Rather than discussing truces and armistices where neither side was a clear winner, and wondering why they do not become permanent (Lebanon supplies a very sad series of examples), we will rather limit ourselves to those situations which history and the immediate commentary would indeed have labelled as a clear victory for one of the sides, and as a concession of defeat by the other. By a certain logic, such a clear outcome should then also be unambiguous in its after-effects, perhaps as clear as when Carthage became part of Rome, or when the United States was recognized as independent by Britain. Yet the model of some of the surrenders since the beginning of this century, especially the Bolshevik surrender at Brest-Litovsk, and then even the German submission at Compiègne and Versailles, suggests that inventive minds have now found ways of challenging such clarity.

But such inventive minds then open many questions. If the winners must now know that the surrendering losers may have their tongues in cheek, have the intention of striking back again when the opportunity arises, i.e. are only crying 'uncle' for the moment – in some deceitful attempt to avoid the immediate suffering being imposed by the winner's forces, why would the winner's commanders then submit to such duplicity? Why accept a white flag of surrender, if the other side is not really going to surrender?

When the Bolsheviks had the temerity to announce to the Germans, at the early rounds of the 1917 Brest-Litovsk negotiations, that this was their intention, the Germans were understandably outraged. What kind of gentleman declares that he is terminating the war by refusing to fight on the next day, but then also reserves the right to begin fighting on another day, or by other means?

Recovering from their indignation, the Germans hit logically enough upon one possible response, ordering their forces to advance in face of this Bolshevik 'no war, no peace' posture, *seizing territory*, capturing assets, imposing additional discomfort on the Russians, i.e. reminding the Bolsheviks that surrender involved a concession on both sides, since the winner normally stops his soldiers from shooting as his part of accepting a genuine surrender. Seeing that they thus indeed needed something from the Germans, the Bolsheviks in the end agreed to sign a surrender that had some extremely onerous territorial terms written into it; but they then at the same time avoided signifying that they would feel bound in any meaningful way by such terms, any longer than military realities dictated (Deutscher, 1970).

By thus adopting postures by which any and all of their promises might be meaningless, the Bolsheviks still confronted the German High Command with an impossibility of extracting the normal gains of a full surrender. This may indeed be a clever bargaining tactic, to show one's adversary that his fondest dream is impossible, for he may then settle for something less, rather than prolonging the mutually painful contest of wills. But what if he is not so inclined? Is peace then impossible?

It is in this realm of the credibility and viability of promises, and the longer-term musterings of energy and hostility affecting the nations facing each other, that the real dynamics of a 'normal' surrender and an abnormal surrender are thus displayed. A normal nation at the beginning of the twentieth century might have been so accustomed to keeping its promises that the new commitments implicit in a surrender would have been quite constricting and meaningful. But this novel Soviet government would for a time free itself of all that was normal and advantageous or constricting in such an attitude. Indeed the Bolsheviks, at all stages, announced that they had no intention of abiding by the terms they were signing, not even having wanted any terms signed at all at the outset, but trying to deliver a 'unilateral surrender'.

Such a unilateral surrender might seem to violate Paul Kecskemeti's general rule, in his 1958 seminal work *Strategic Surrender* (Kecskemeti, 1958), that the loser always gains something in exchange for his act of ceasing to man his defences. Just as there could be nothing like a truly 'unconditional surrender' in Kecskemeti's analysis, there could not be anything like a unilateral surrender, for there would be an implicit moral injunction on the winner to himself cease firing (and also to cease advancing?), for which the loser must also offer something in exchange.

Yet the unilateral surrender gambit implied no Bolshevik promise to abstain from resuming hostilities in the near future, and no Bolshevik acceptance of the legitimacy of any new status quo. The latter Bolshevik concession was then to sign a surrender document, but to refuse to discuss it or to haggle about it, i.e. to be willing to sign anything, and thus to try to take the legitimacy and commitment of faith out of it as much as possible, striving to render the document as signed comparable to no document at all.

A novel tactical approach to the offering of one's word thus makes a tremendous difference for the contractual process by which wars are ended. But perhaps there was a second difference, in that the 1917 victory, and the German demands, were themselves of proportions that were abnormal. The Bolsheviks, faced with the totality of their military defeat by the Germans, could well have rationalized to themselves that this totality would have to be but a passing phase in the flow of international relations, such that any peace simply reflecting the situation of 1917 would be subject to revision. Regardless of what terms were signed, therefore, regardless of what promises were made, these would all have come 'under duress'.

A traditional analyst of diplomacy might still scoff at the duplicity of someone signing a peace treaty which he intends to overturn at first opportunity, responding that *all* international dealings are influenced by 'duress', as power relationships are the input that determine the outputs of such things as boundaries or peace terms. Yet there clearly is some distinction between obligations which are entered into in the context of very *unusual* power situations, and those which are entered into with a view to permanence and compliance.

Such a distinction may become all too valid in the future for situations of hostage-takings, and of other terrorists acts which place a great deal of civilian life at risk. States may have to make promises to free such hostages; but will states feel obligated by such promises thereafter, or will such states not feel very free to behave in ways very counter to any promises that were exacted while the hostages were still being held, working to punish and strike back at the terrorists involved (Mickolus, 1976)?

As an immediate example of how the Bolshevik example could be emulated, some of the German advocates of signing the Versailles Treaty in 1919 argued that it should be treated in exactly the same way (Wheeler-Bennett, 1939): the signature would be delivered in as speedy and perfunctory a manner as possible, thus to set the stage for as early a revision as possible. Everyone in Europe wanted to get the war over with in 1918. If the Allies would be pleased by drawing boundaries capriciously, and by having Germany accept 'war guilt', let them have this on paper, as long as the Germans understood amongst themselves that this guilt was not something they were to feel, and as long as the boundaries would be challenged politically or militarily as soon as Germany recovered from its temporary disadvantages.

Other examples

On the issues of the 'verbal surrenders' here, we could point to many analogous examples in our own practice. The Communist regimes insist that we refer to East Germany as the 'German Democratic Republic', and to China as the 'People's Republic of China', with the takeover of the bulk of China by Mao Zedong's forces being referred to over and over again as the 'liberation'. If Western liberals find themselves using such phraseology in dialogues with representatives of these regimes, are they thereby surrendering their honest beliefs that East Germany is not democratic, or that the People's Republic of China has largely been ruled by individuals quite willing to overrule the preferences of most of 'the people'?

Many of us typically condition our thought processes so that we are not at all making such a surrender, as we mentally put quotations around the phrases we use, voicing such phrases under an effective duress – that these Communist regimes indeed do have control over these areas, so that we can not visit Leipzig or Shanghai without a visa from such regimes. 'Call yourself anything you want to, and we will echo the nomenclature, simply as a concession to your political power' – this would be our mental response to this enforced kind of dialogue. By putting political power behind what purports to be a descriptive adjective, the regime in question takes all the meaning out of that adjective.

Yet the game is not totally settled here either. There are indeed Western liberals who worry that we can be brainwashing ourselves in giving in on nomenclature. There were similarly many Germans in 1918 who worried about what would be done to German memories and German attitudes about moral issues, if the 'war guilt' finding of the Versailles Treaty were accepted. The issues of nomenclature may fade somewhat in importance when compared with arrangements of boundaries and strategic positions and industrial resources and future force levels, but they open all the same 'gaming' issues we will be discussing throughout this article. Here, as elsewhere, there are cases where winners can make permanent what they feel they have accomplished at the moment of the 'surrender', and there are other cases where this is overturned quickly. The possible lessons of how one is to make the verbal concessions permanent are of great importance and value to the winners (Oren, 1982). The lessons on how one can make them transitory and impermanent are similarly of importance to the losers. Material considerations play an important role, but moral attitudes and expectations fill out the rest of the equations.

For a still-earlier example of a side 'surrendering' without any intention of really relinquishing what was being contested, one need turn only to the American South at the end of the Civil War. As Grant let Lee keep his sword and let Confederate soldiers return home to their farms, the Southern states were busily devising ways of keeping the substance of black slavery and political subordination in place, despite the wording of the new Thirteenth Amendment to the US Constitution (Carter, 1959).

The ensuing struggle is also interesting for what it might show about our general theme of the permanence or impermanence of military victories. The Southern states seemed by the winter of 1865–66 to be thumbing their noses at their victorious adversaries, as the 'Black Codes' were enacted. The reaction of the Republicans in the House and Senate was not to give in, in face of this, but to overrule the accommodating inclinations of President Andrew Johnson, and to impose a more severe military occupation in the form of Reconstruction. The South's response then came in a mixture of patience and minimal compliance, punctuated with the guerrilla warfare activities of the early Ku Klux Klan. By 1876, the entire picture looked more like a defeat for the abolitionists in the North, as Reconstruction was terminated and federal armed forces were withdrawn from the South, and as black voting rights were withdrawn by one strategem or another across the states of the former Confederacy, with the segregation of 'Jim Crow' laws being imposed thereafter.

Yet the story does not end there. The struggles for black freedom which had seemed so lost between 1876 and 1914 were resumed during and after World War II, with the civil rights movement in the end restoring the rights of black America to vote all across the South. Did the victories of Federal forces in the Civil War, by which the political union of the geographical United States was restored, thus come to mean nothing by the time of the first US Centennial celebrations in 1876? Or was this victory quite real, but merely one of delayed action? With the United States tied together by the economics and laws of a single continental union, it may have been inevitable that the rights enjoyed by black citizens in the North would sooner or later have to be accepted in the South as well.

Earlier questionings of 'victory'

Those who surrender are hardly always the winners in the end. Paul Kecskemeti's *Strategic Surrender* argued quite logically that no surrender was ever unconditional, in that the loser was agreeing to cease firing whatever bullets he had left, and thus would have to be offered something in exchange for this concession, i.e. that the exchange, however one-sided, still also involved some 'conditions' imposed on the victor. Kecskemeti's analysis of the French surrender to Germany in 1940, and then the Italian, German and Japanese surrenders to the Allies in 1943 and 1945, showed that the losers were able to extract a number of important provisions for the future, even while they were making the future of the victors more attractive by agreeing to lay down their arms.

Yet Kecskemeti's analysis may have proved too much, in its stress on the terms accepted by the victors, just as our normal intuitions may have projected too much in the opposite direction, in assuming that the losers were committed forward into posterity to whatever terms they had agreed to. What is fascinating in the analysis of differing surrender arrangements is how tongues in cheek sometimes become very relevant in the next decade of two, and in other cases quickly lose their options. The terms of a surrender and armistice are agreed to because each side is tired of fighting, but what is it that breathes life into such terms thereafter, that leaves neither side pining for a revision or a double-cross (Oren, 1982; Fox, 1970, Ikle, 1971)? This issue of 'real peace' related to the dynamics and honesty of motivations for both the losers and the victors.

There is another school of analysis, less analytical than Kecskemeti's, and more intuitively inclined to see higher moral patterns which would see victories and surrenders as inherently subject to revision. Only a negotiated truce, with neither side identified as the victor or loser, can produce an enduring peace or 'real peace', one that both sides are ready to internalise and accept (Keynes, 1920). Yet this kind of analysis would not be able to explain Mexico's acceptance of the Rio Grande border, or Japan's resignation to the loss of Taiwan and Korea, or any other successful war termination which has not simply returned the powers back to the *status quo ante*.

Not every 'fair peace' becomes an enduring peace, and not every 'winner's peace' has to be impermanent. Revanchism may be produced by a victor over-reaching himself, but even this has to be hemmed in with qualifying conditions.

The role of moral attitudes and expectations

What we are dealing with here is a lack of continuity in the keeping of promises ending a war, the promises of a surrender. Whether promises must be kept, or can be ignored, after signing a capitulation, will depend on the opinions of the outside world. Some of this tendency to attach no importance to a pledge has been a tactic deliberately and callously cultivated by innovative political forces. But some of this tendency has also sometimes been the result of secular changes in the natural background of the situation, as power relationships now often appear more transitory and temporary. Much will depend on how the outside world compares these sources.

If someone can take all the meaning out of a signed confession, or a surrender, or a peace settlement, what can the other side then aspire to? What is the point of winning if the defeated side will raise its hands in surrender, and then withhold its fire only for the moment, beginning its revisionist counter-strokes at the earliest opportunity? We keep promises, Frederick the Great commented, because otherwise no one will trust our promises in the future. We respect the terms of a surrender, for fear that no one will let us surrender in the future, not trusting our intentions when we run up a white flag, instead of continuing to mow us down until none of us are left.

Where this world would normally be intolerant of the false surrender tactic, however, it will be more tolerant where the conditions of duress are made to look very temporary and 'abnormal'. Would this outside world thus think so badly of President Reagan and the United States if retaliation was directed at the terrorists who seized a TWA flight, even after a promise had been made that no revenge would be sought? Or would the world tolerate and forgive any such violations of promises on the US side, not losing faith thereby in the American tendency to keep promises, because the implicit or explicit pledges conveyed to terrorists were the product of a very abnormal and transitory power relationship?

Total war presses the conventions of international behaviour to their limits. This is hardly a new observation. Yet one concomitant of what I have outlined here is that the conventions of war termination will thereby fall more and more into question (Cimbala, 1986). A country which under other circumstances would have drawn world distrust and contempt when it mocked its capitulations to a military superior will be forgiven this, if this situation of military superiority is carried to an extreme, if peace terms seem to be imposed from some kind of total, but thereby inherently temporary, dominance.

Obviously, therefore, any holder of momentary military advantage must now continually contemplate how to guide that advantage for the future, and this may be more concretely what the advocates of a 'just peace' or 'real peace' thus have had in mind, as they urge their governments to play the role of 'generous victor'. This is the problem of Palestinian terrorists, or of the German High Command in 1917, or of the Allies in 1918, or of Mrs Thatcher after the surrender of the Argentinian forces in the Falkland Islands; it has been the problem for the Israelis after each of their military defeats of Arab armies. How does one get the losers to internalize and respect the commitments that have just been extracted at the end of the violent exchange? How does one keep revisionism and revanche from overwhelming sentiments of promise-keeping and commitment?

The role of geographic factors

For natural reasons, and because of the tactical inventiveness of some states, the world has now become more accustomed to the deliberate breaking of peace terms. The task here is to sort out how some use can still be made of the process of surrenders in the termination of wars.

When the German High Command in 1917 found itself being manipulated by the ways that the Bolsheviks used words and promises, they overcame their frustration by seizing territory, by taking a more tangible grip on the future. Yet such geographical approaches can also be uneven in their effectiveness.

Some might assume that the geographical aspects of the settlement of any war will always have far more impact on the future of a peace than the seemingly more trivial issues of whether one accepts the enemy's wordings or not. The 'war guilt' phraseology might sink in, if it were taught as such in German schools, which was not the case after 1919 as it was after 1945. But one design for new boundaries could inhibit future German military challenges or trim back German appetites, just as other boundaries could increase the capabilities of German military and political forces or could whet such appetites.

The various French proposals for dismembering Germany further in 1919 (Nicholson, 1933) by supporting a 'Separatist' independence for the Rhineland, or by actually advancing the French and Belgian borders to the Rhine, clearly thus had in mind the risks and possibilities of such a German 'incomplete surrender'. Yet the attitude of the United States under Woodrow Wilson (who was most concerned that the peace be a 'just peace', i.e. a peace accepted on its merits by the losers as well as by the winners), and of the British under David Lloyd George (who were a little bit more concerned that France not become too powerful on the European continent in the process of weakening Germany – this being the logic of classic balance-of-power mechanism) had precluded the drawing of boundaries that would strip Germany more generally of its ability to challenge the new status quo. The demilitarization of the Rhineland, together with the occupation of this territory by Allied forces for a term of 15 years, was more 'realistically' (i.e. more in accord with French reasoning) intended to hold down German capabilities for revision, while such desires for revision might cool. It is a generally plausible proposition that the taste for irredentism and revision of boundaries diminishes over time, but the lengths of time that this takes can be extraordinarily long, with the Middle East being only one of the areas that supply depressing examples of such irradicable irredentist memories. The most tangible gains that a victor can seize in the process of surrender might thus be the geographical positions that thereafter enhance his military potential. The tops of hills have always been useful for dominating the lands they overlook, with the Golan Heights supplying a very contemporary example. There is such a thing as territorial position, in terms of elevation, or accessibility, that strengthens the defence, such that any future attacker will have to outnumber the defenders by much more than the legendary three-to-one margin. If the price for accepting the ceasefire momentarily desired by the other side (even if it is a side intent on reneging and resuming hostilities at the first opportunity) were thus to be the possession of such bastions and positions of advantage, then as winners we would retain something more permanent as the gains of our adversary's surrender.

One general rule would seem to apply: the more an adversary has taken meaning out of his pledges, the more one would have to seize territory as a way of guaranteeing an end to hostilities. But some of such territorial advances could instead increase vulnerability to counter-attack, as we were overextended in the process and more prone to ambush, or as we had more difficulty in supplying and reinforcing advanced salients. When the Germans in 1916 withdrew from a portion of the territory they had been holding in France, this actually enhanced their ability to hold on to the rest (Liddell-Hart, 1930).

The taking over of advanced territorial positions obviously also has ramifications beyond its contributions to, or burdens to, the strength of the defence. Considerations of offence and defence in this more narrow and precise sense apply entirely to the counterforce aspect of warfare, whether one army can hope to defeat another army in a future war, which is classically only one half of what war is all about. But the other, 'countervalue', aspect of warfare must also be constantly kept in mind, i.e. what the changes of territorial control in the wake of a surrender can do to each side's ability to impose pain and disutility on each other, and what such shifts of territory may already amount to, in terms of an immediate and continuous imposition of such pain.

Countervalue warfare is very much about motivating the other side, about getting him to change his preferences, getting him to forego using his military capabilities. To motivate our opponent, to deter him, we must be in a position to impose punishment if he takes military action against us, and to withhold such punishment if he withholds such action.

Staking out an advanced territorial position, as when the Germans took over all of Alsace-Lorraine after 1871, or when the Israelis assumed control over the West Bank territories after 1967, may thus amount to a permanent (or at least long-term, until the other side writes off the territories lost, and becomes inured to the 'pain' of having lost them) imposition of disutility on the other side, withdrawing one of the cards with which one could have bargained. Staking out such an advanced position may also open one to countervalue attacks from the other side, especially if one has not merely implanted a military presence in the most appropriate locations of such territories, but has also brought along civilians dependents, settlers, or colonists. Geography thus may supply some counters and answers to the problem posed by duplicitous surrenders, but the solutions offered here will have to be quite complicated. The seizure of territory sometimes offers ways of making the other side's surrender more meaningful, but only sometimes.

Some conclusions: the difficulty of playing one's hand

One of the more difficult tasks of judgement thus comes in assessing how and when the side which lost the last round of warfare can be brought around to accepting that loss, i.e. how the second side can be brought around to endorsing the international status quo with which the first is already quite content. In an earlier day, this was called 'appeasement', when this phrase was in no sense something so pejorative, so as to be ridiculed or deplored. The ways that Hitler and Mussolini then exploited such processes and sentiments, with Hitler asking for so much to redress the alleged incompleteness of justice at Versailles, that he had troops in Paris and at the gates of Moscow and Leningrad at the end, has forced us not at least to use a different phrase, since the associations of the 'appeasement' phrase seem irradicably negative. Yet the lessons of the endings of World War I and World War II are hardly that either a total dominance and vindictiveness ('the Carthaginian peace'), or a total magnanimity, are the avenues to desired solutions (Gilbert, 1963).

'Incomplete surrender' is in part a natural development of the nature of international power relationships, but in part also a clever artificial device. We may all feel more fond of the time when men were as good as their word, when cries of 'peace' and 'surrender' and 'Kamerad' meant that the other side was really prepared to stop the shooting. Yet this time may always have been just a special time, special in the nature and stability of nations, and in the stability and regularity of international relationships.

References

Carter, Hodding (1959). *The Angry Scar*. Garden City, New York: Doubleday.
Cimbala, Stephen J. (ed.) (1986). *Strategic War Termination*. New York: Praeger.
Conquest, Robert (1968). *The Great Terror*. New York: Macmillan.
Deutscher, Isaac (1970). *The Prophet Armed*. New York: Oxford University Press.
Fox, William T. R. (ed.) (1970). How Wars End, *Annals of the American Academy of Political and Social Science*, No. 392, November.
Gilbert, Martin (1963). *The Appeasers*. Boston: Houghton Mifflin.
Handel, Michael (1982). War Termination: A Critical Survey. In Nissan Oren (ed.), *Termination of Wars*. Jerusalem: Magnes Press, pp. 40–71.

Harkabi, Yehoshaphat (1972). *Arab Attitudes Toward Israel*. Jerusalem: Israel Universities Press.

Ikle, Fred (1971). *Every War Must End*. New York: Columbia University Press.

Kecskemeti, Paul (1958). *Strategic Surrender*. Palo Alto: Stanford University Press.

Keynes, John Maynard (1920). *The Economic Consequences of the Peace*. New York: Harcourt Brace.

Liddell-Hart, Basil H. (1930). *The Real War*. Boston: Little Brown.

Mickolus, Edward F. (1976). Negotiating for Hostages: A Policy Dilemma, *Orbis*, XIX (4), 1309–25.

Nicholson, Harold (1933). *Peacemaking 1919*. New York: Grosset and Dunlap.

Oren, Nissan (1982). Prudence in Victory in Nissan Oren (ed.), *Termination of Wars*. Jerusalem: Magnes Press, pp. 147–63.

Rosenberg, Arthur (1959). *The Birth of the German Republic*. New York: Doubleday.

Wheeler-Bennett, John W. (1939). *The Forgotten Peace*. New York: Morrow.

6 If Cold War is the Problem, Is Hot Peace the Solution?

Peter J. Taylor

> If peace is based on freedom, then the struggles for freedom must become the first concern of the peace movement if there is to be any escape from the web of nuclear blackmail spun by the superpowers. . . . The peace movement will only survive and grow if it extends its campaign against armaments into support and encouragement of every expression of independence from the two dominant powers. . . . In order to become successful as a peace movement, the movement also has to become a libertarian movement (Migone, 1987: 63–4).

Towards the end of 1947 Walter Lippmann's articles refuting George Kennan's new containment thesis were brought together in a book Lippmann chose to call *The Cold War* (Steel, 1980: 445). The phrase struck a chord in the public consciousness and ever since we have been living in the 'Cold War era'. Lippmann's initial success in making such a spectacular addition to our political lexicon was due to the profound aptness of this phrase. Neither war nor peace, 'cold war' cut through the confusion to define a highly militarized 'peace' of threat and counter-threat.

The longer-term success of Lippmann's phrase was due to the particular nature of the post-World War II world. The continuation of peace and war threats was certainly not unique historically. What was new was the perception that world politics had now become fundamentally ideological. Cold War came to represent a fractured world where two sides were stuck seemingly in antagonistic postures for ever more. As Edward Thompson (1987: 14) has portrayed it, international relations seemed to have become 'glaciated'. Such physical analogies mark an era where threats of war and the need for military preparedness have made a mockery of all hopes for truly peaceful change. As the Cold War itself retreats into history we can be sure that there will be little nostalgia for its discipline and terror.

But the phrase Cold War remains a fascinating one. It encapsulates many of the contradictions inherent in peace studies. It is a 'peace' – Gaddis's (1987) long peace between the superpowers – that is called a 'war'. The confusion between the goal of studying peace but the necessity for studying war seems to be a conundrum from which political geographers have had trouble escaping (Pepper, 1985). The notion of Cold War compounds their difficulties. For instance, what is the opposite of cold war? Presumably Lippmann intended his phrase to contrast with the hot war with Germany that had preceded this cold war. But the climatic analogy does not stand up well to more recent interpretations, as Halliday (1983) points out. Nowadays a warming up of the situation or 'thaw' leads to improved international relations, a *detente*. Hence if the public is informed that world politics is 'blowing hot and cold' they can be excused for not understanding which phase, hot or cold, is the belligerent one!

In short, cold war provides an ideal seed for exploring the ambiguities of peace as a political goal. Drawing on the European peace movement literature, I shall suggest a way out of the political geography *cul de sac* in peace studies. As will have been noted from the title of this chapter, I will posit 'hot peace' as a viable and laudatory opposite of cold war.

My route to this conclusion involves a discussion of Fernand Braudel's social time concepts. This represents the theoretical structure of my argument. These concepts are then applied to a world-systems analysis interpretation of wars and their subsequent peaces in which our recent Cold War can be located. Armed with theory and fact we can move on to explore hot peace as a future alternative to cold war that is worth constructing. Finally I return to political geographers by way of a conclusion to consider the contradictions in their roles as world citizens and professional intellectuals.

A three peace suite

The reason why peace is a problem for analysis is that it is often treated as simply the opposite of war. As a negative category it is difficult to theorize about. Clearly the concept of 'not-war' leaves a lot to be desired. But by using this bald definition we can begin to see a way forward. By defining it as the non-existence of war, the concept of peace is a prisoner of the history of events. War breaks out, shattering the prior peace; war comes to an end, peace returns and so on. Fortunately this is not the only kind of history that exists.

Fernand Braudel (1980) has argued against the dominance of 'eventism' in conventional history. He identifies three 'social times' that relate to the way in which societies are constitutional through time. He is most famous for his *longue durée* concept, the very long-term trends or structures that underpin all societies. These are expressed in the day-to-day activities of the population that continue regardless of the more rapid political, social, and economic changes going on around them. Braudel is referring here to cultural histories or even 'geographical history'. At the other end of his social time there is a history of events, *histoire événementielle* or *courte durée*. This orthodox history is typically political in its bias and deals with change in terms of particular episodes. In between the *moyenne durée* is often the time approach adopted by social and economic historians with their interests in various cyclical patterns of change. For a complete history, a full understanding of social change, Braudel argues that we need to take into account all three social times. Like geographical scales of analysis, these historical spans are not autonomous; all three exist together as part of a single, historical whole.

What does this all mean for the concept of peace? It is simply that we can identify three kinds of peace, each associated with a different social time span. The peace of the *court durée* is derived from the urgent concern to find mechanisms to stop war. It consists of cease-fire provisions, armistices, peace-keeping forces and other immediate tools against violence. Alternatively 'victory' for one side immediately creates a peace. This is the simple idea of peace as 'not-war'. The peace of the *moyenne durée* is a constructed peace, the product of a negotiation that sets up institutions to maintain a political order. In this sense it is an imposed peace, of victor over loser, and its main purpose is to create a stability where there had been disorder. Such 'stable peaces' need not be literally 'not-war' since minor wars are tolerated as long as they do not threaten the overall political order. The Concert of Europe after 1815 and the League of Nations after 1919 are examples of such medium-term peace institutions but with contrasting fortunes. The peace of the *longue durée* is altogether different. This is a peace whereby the fundamental structures of the social system negate violence. In this peace the bases of war that exist in all exploitative class societies are removed. The logic of the system no longer rewards violence as a means of social change.

It is argued here that any comprehensive peace studies should accommodate all three 'peaces' into their analysis. Part of the problem has been, however, that each type of peace has been viewed separately. This has been exacerbated by the various political associations that each peace has attracted. For instance, the peace of the *courte durée* has been dominated by liberal political concerns to put an immediate end to existing violence. As an example, we can note that this dimension of UN business has been the particular concern of Scandinavian countries. Although the peace institutions of the *moyenne durée* in the twentieth century have had similar liberal associations, we can argue more generally that stable peaces and the resulting maintenance of a status quo are inherently conservative. The concern for order and stability are arch-conservative values and they produce a peace which we call pacification. The peace of the *longue durée* is, of course, transformational. It is typically the peace of a radical politics in its requirement for fundamental change. But we must not take these political tendencies too far. Any politics of peace most incorporate all three time spans to have a complete theory of peace. All too often, of course, political discourses get stuck in one or other time span – hence the political associations with peaces already noted.

We can illustrate the advantages of applying Braudel's social times to the analysis of peace by reviewing a recent debate in political geography over peace studies. Van der Wusten and O'Loughlin (1986) invited political geographers to cease neglecting peace issues by providing them with an agenda for 'war/peace studies'. The new agenda turns out to be new in content (peace) but orthodox in method (positivist). The positivist approach has been attacked by O'Tuathail (1987) as an instrumentalist problem-solving model of science. Basically O'Tuathail interprets van der Wusten and O'Loughlin's agenda as inviting further exercises in *realpolitik* in searching for stable foreign policies. In contrast O'Tuathail calls for a critical science approach that questions the social and political relations of contemporary world politics. Van der Wusten and O'Loughlin's (1987) rejoinder criticizes O'Tuathail for providing no agenda to deal with the practical issues of the day. What does this all mean in Braudellian terms? Clearly O'Tuathail's peace is a phenomenon of the *longue durée* whereas van der Wusten and O'Loughlin are concerned with the *court durée* and the *moyenne durée*. Both seem to me to be correct in their criticisms of each other. Van der Wusten and O'Loughlin's search for 'stable peace' cannot begin to find a way to a more general long-term peace without querying the nature of our current material society but O'Tuathail's radical critique is likewise in need of what is called 'non-reformist reform' in other contexts, that is to identify and promote short-term and medium-term changes that are consistent with the ultimate longer-term peace goals.

Arguing that all three peaces are necessary in our analysis does not mean that they must be treated as being of equal importance. This discussion is closer in spirit to O'Tuathail than to van der Wusten and O'Loughlin because I take *longue durée* concepts to be ultimately the most important. Although other peaces have their place, our sights should always remain set on questions of the *longue durée*. In this peace study, therefore, we start with the long term and work our way back towards the events that make our political world.

Historical peace and perpetual peace

There is a fundamental difference between peace defined in *longue durée* terms and the other two types of peace. When we consider peace from either *courte durée* or *moyenne durée* perspectives we can provide concrete examples of what we mean. We are arguing in the realm of historical peace. In contrast there has never been a *longue durée* peace. We are arguing in the realm of what can be, such as Kant's 'perpetual peace' or Galtung's 'true worlds' (Gallie, 1978; Galtung, 1980). Such preferred worlds have never existed so that we are inevitably utopian in our discussion. But that does not mean that we must be divorced from the realities of history.

Historical peaces are stable peaces. They encompass a status quo in which there are winners and losers. In this sense all peace settlements, not just Versailles, contain the seeds of the next war. Such peace has been the concern of political geographers from Versailles onwards through to van der Wusten and O'Loughlin's 'claiming new territory for a stable peace'. The problem is that unless conditions exist for a voluntary peace, all such stability will ultimately have to be imposed. And it is when we come to consider these necessary coercive conditions that we are forced to move beyond the narrow political confines of such peace studies. Before we can even contemplate a perpetual peace we must broaden our understanding of the limitations of historical peace.

For Johan Galtung (1980) the contemporary world is simultaneously facing four crises relating to violence, misery, repression and environment. Each crisis has generated specific politics with the goals of peace, welfare, human rights and ecological balance. The movements behind these politics all have programmes to reach their goals but they are all doomed to failure, according to Galtung, unless they recognize that the particular crises they tackle each have their roots in a single world structure. Hence there is no long-term benefit to be obtained in promoting and adopting a solution to any one crisis if the other crises are not simultaneously addressed. In Braudellian terms all four crises are phenomena of the *longue durée*: perpetual peace necessitates welfare, human rights and ecological balance. These are the conditions that make our preferred peace fundamentally transformative in nature.

The political implications are, of course, profound. As well as directing policy against actors as perpetuators of violence, one must also target the structures that are equally implicated in violence. In fact, it is relatively easy to show that the latter is far more important by simply quantifying violence in terms of preventable deaths (Johnston *et al.*, 1987). Violence-as-action or direct violence, whether state-sanctioned as in wars and death squad 'disappearances' or personal attacks as in murder or manslaughter, generally numbers its victims in tens of thousands per annum. Violence-as-structure or indirect violence, namely, those who die unnecessarily, consistently numbers its victims in tens of millions per annum. Hence the politics of peace must have two elements; to combat direct violence there is the quest for security and to combat indirect violence there is the fight against domination or the struggle for liberation. These two politics have been recognized in countervailing violence in the twentieth century. Galtung identifies the classic examples in World War I as 'the war to end all wars' and the Russian Revolution as 'the domination to end all domination' (i.e. dictatorship of the proletariat). These solutions may have failed but the logic endures.

Perpetual peace assumes the removal of all possible impediments to individual and group self-realization. In our previous terminology it is 'not-war' plus conditions for not-war. Such a utopia implies an unstable pluralism to match the variety that is humanity. Instead of a stability that is 'cold' we can picture a mosaic that is 'hot' and therefore alive. But before defining such a world, we must return to our own long peace and locate the Cold War among historical peaces.

Locating the Cold War

The Cold War was proclaimed and maintained in *longue durée* terms. In the late 1940s both camps used a language that viewed the conflict as the ultimate test of humanity, either as global class conflict or as liberal democracy versus totalitarianism. In President Truman's words two alternative ways of life were on offer to the world. In short we were presented with a great clash of 'civilizations' (Taylor, 1990).

As a generally perceived contest between opposing structures, the Cold War was different from all other historical peaces. But as we come to the end of this era such language sounds rather hollow. The presentation and the reality seem to be very different. Although some cold warriors insist on welcoming Eastern Europe 'back to civilization' (Hawkes, 1989), such terminology is no longer so generally in vogue. Now the Cold War looks very much like previous peaces, a point emphasized by John Gaddis's (1987) description of it as a 'long peace' to mimic the more well-known nineteenth-century predecessor. In political geography terms the Cold War is not a phenomenon of 'civilizations' or structures, rather it is simply a geopolitical world order (Taylor, 1989, 1990). It is an example of inter-state relations which form a distinctive global politics. As such it is one of several world orders that have successively formed the political superstructure of the modern world-system. But it is one of a fairly rare kind of world order that is associated with the peace at the end of global wars. Viewed this way the Cold War is not unique but it is a special kind of world order.

In world-systems the Cold War geopolitical world order is a particular phase of a hegemonic cycle. These cycles of a century or more encompass the successful rise and the failing demise that surround a period when one state is able to dominate the system economically, politically and culturally. There have been three such hegemonic states – the Netherlands, Britain and the USA – whose cycles centred on the mid-seventeenth, mid-nineteenth and mid-twentieth centuries respectively. In each case a global war of approximately 30 years resulted in the defeat of the main land power challenger, leaving the hegemonic state to reap the spoils. In the original Thirty Year War the Hapsburg challenge to the Dutch was finally ended; the Napoleonic Wars confirmed British leadership with the defeat of the French; and the World Wars I and II, 'the German wars', ensured that the USA rather than Germany would be Britain's successor. Let us look at the imposed peaces that followed from these global wars and which resulted in hegemonic states.

After the disorder and uncertainty of thirty-year global wars it is inevitable that order and stability dominate the world political agenda. The religious strife of the original Thirty Year War was countered by the provisions of the Treaty of Westphalia (1648) which is usually interpreted as the formal origins of international law (Gottmann, 1973). The religion of the monarch was deemed to be the religion of the state but most important in the longer term, was the internal sovereignty of territorially defined states that this implied. Non-interference in the internal affairs of other countries became the first law of international relations. This allowed for the pacification of territories by central state structures. This first post-global war peace settlement, therefore, was achieved at the scale of the state.

This left untouched inter-state relations which were perennially warlike in the competitive mercantilist era culminating in the Revolutionary and Napoleonic wars. It was at this time of perpetual wars that Kant in 1795 produced his proposals for an inter-state peace in a voluntary confederal arrangement which he termed perpetual peace (Gallie, 1978). In the event, the peace that followed this second thirty-year war was another imposed peace but on a much larger geographical scale. The international system known as the Concert of Europe was based upon a relatively even balance of power among a few large states that guaranteed the peace (Langhorne, 1981). The result was the European 'long peace' from 1815 to 1914. The scope of this arrangement was defined specifically as Europe's leaving Britain a free hand in the southern continents and Russia a free hand in central Asia. This was pacification of Europe by imposing a great power order leaving the rest of the world with the more traditional mixture of war and peace. With changes in the balance of power, notably the rise of Germany, and new communications leading to a global broadening of important international relations, particularly into the Far East, the Concert of Europe gradually collapsed. With the end of great power pacification of Europe came the next thirty years war, the German wars of the twentieth century.

We can now locate the Cold War as the outcome of the third global thirty years war. It is, therefore, the third great 'freeze' imposing a stable order on part of the modern world-system. In this case the zone of pacification encompassed the whole of the 'North' from North America to Japan and including Europe. This vast zone became frozen into two antagonistic camps. Beyond the stable peace region the traditional mixture of war and peace was to be found once again. For instance countries were allowed to 'change sides' in the 'Third World' where they did not threaten the overall world order while the 'first' and 'second' worlds remained frozen. There seemed to be an unwritten international law prohibiting serious encouragement of 'swapping sides' in the 'North'.

The Cold War geopolitical world order and its 'long peace', therefore, fits into a pattern of increasing scale of pacification in world-system hegemonic cycles. For proponents of 'stable peace' this may be considered heartening as an island of peace that we may be able to diffuse to other areas (Boulding, 1978). But in reality what has been achieved by these historical peaces is a relatively peaceful core leaving the periphery as unstable and disorderly as ever. As the world-system has grown so the core has increased in size, enabling the construction of even larger islands of stable peace within successive hegemonic cycles while the periphery as a great 'sea of disorder' has also grown with the system at the same time, with the result that the overall balance of order and disorder may not be very different. Hence in place of an optimistic peace diffusion model, a world-systems analysis provides a pessimistic, structurally balanced, order–disorder mode. Historical peaces of the *moyenne durée*, of which the Cold War is one, are very partial peaces.

But the Cold War is also a very distinctive 'long peace'. Its programme of pacification is quite unlike any other. There are three unique features. First the hegemonic power, the USA, has been unable to translate its world economic dominance into a global political and ideological dominance. When Britain was hegemonic, for instance, there was no equivalent of the USSR political challenge. Second, there is another unique challenge. The non-aligned movement in the 'Third World' is a large collection of states that profess not to choose sides in the Cold War. They refuse to accept the organizing principle or this geopolitical world order. Third, and most important, this world order is cemented by an ideological conflict out of all proportion to previous world order. For instance, in 1914, the leading liberal state, Britain, entered World War I in alliance with the arch-conservative state, Czarist Russia. In contrast the Cold War was moulded in *longue durée* terms. Its stability has been based on competitive pacifications. In this sense it resembles more the 'peace' of the age of mercantilism, with its competitive state territory pacifications, rather than the single zone pacification of the long peace of the nineteenth century. What the competitiveness does reveal is the pacification inherent in any imposed world order. This points us beyond the *moyenne durée* and places the peace of the *longue durée* firmly on the political agenda. From our experience of the 'long peace' of Cold War we can appreciate the necessity for a non-pacification peace, 'perpetual peace', 'true worlds' or our 'hot peace'.

Hot peace for Europe

The most telling indictment of the way in which Cold War pacification destroys diversity has come from the pen of Caesar Voute (1987) in his evocation of a past 'cultural' Europe. It is entirely appropriate that we should begin our search for hot peace in Europe, not for traditional Eurocentric reasons, but because this is the continent in which the Cold War was constructed. It can be argued, therefore, that this is the place where our bi-polar world must be challenged first. Furthermore it is Europe that has bequeathed to today's world the modern practice of pacification by territory. If this can be undone in its European heartland then global hot peace must become a real possiblity.

Voute (1987: 154) argues that to be a European in the 1980s 'is to go through life with an acute sense that something is missing, has been cut off from us'. This is not just a product of the division of Europe preventing Europe-wide intercourse, rather the Cold War is merely the latest symptom of a far older malaise which he describes as 'this feeling of being culturally somehow incomplete' (p. 155). The European culture in question is one of great diversity where different religious, language and ethnic groups live side by side in a geographical mosaic of accommodations. Voute does not pretend that this is a perfect world of harmony – 'Our heritage of creative diversity was always mixed up with intolerance, local racialism, wars and conflict' (p. 157) – but there was a definite cultural vitality that has been eroded by the politics of uniformity. It is this positive aspect of cultural vitality that Voute hopes to reclaim from the Cold War.

He uses a compelling metaphor to make his case: Europe's myriad cultural streams have been gradually diverted into stagnant pools of conformity. In these cold, silent worlds reptiles rule, eliminating any warmth or diversity. The original destructive forces – the first reptiles – were the state nationalisms that converted borderlands of cultural exchange into national boundaries of separation. Thus long before the imposed conformity of the Cold War: 'Abhorrence of untidy diversity and regulation of uniformity had become principles of administration of whole states' (Voute, 1987: 157). Anglicization in Wales, Germanification in Poland and the many other intolerant standardizations of peoples culminated in the national purity programmes of fascist Europe. The great tragedy of Europe is that the defeat of the latter produced an even greater demand for conformity. By 1947 the 'reptiles' were back and 'the healing of Europe' that had begun was lost (p. 159). Hence for Voute 'the supposed entities, "East" and "West" are ideological fictions, products of the reptile house' (p. 163). It is this Europe dominated by two massive stagnant pools that is the Cold War. This new 'supernationalism' based upon 'the permanent enemy hypothesis' (Thompson, 1987: 21) is the contemporary obstacle to direct contact with Europe's cultural diversity.

So what exactly is 'hot peace' ? It most certainly is not a nostalgia for a past harmony before the 'historic nations' of Europe were collected into just a few 'nation-states'. Maps of such 'little nations' (e.g. Williams, 1989: 253) have no historical basis in that Europe has never been a collection of small, self-governing communities. Some historic nations may achieve independence in the future but the renaissance of cultural diversity in Europe cannot be based upon a frozen heritage that never existed. If Europe is to have genuine diversity then the communities will have to be constructed. As Ascherson (1989) has argued, these may be based upon historic nations but other units, administrative and functional, may equally become foci of local consciousness.

Such change cannot be achieved without social conflict. Constructing communities is a continuous process and intercommunal tension is inevitable. Hence this Europe is 'hot'; it is alive. For it to be a 'peace' there must be institutional arrangements to facilitate accommodation without pacification. This will require structural change to work. Here we note that 'hot peace' is the complete opposite of Fukuyama's (1989) end of history contention in which a future of dull conforming societies is predicted. Such an end of history would represent the victory of the reptiles.

One political geography feature that hot peace will require is the return of borderlands to replace precise boundaries. These will not be traditional borderlands of Europe but new zones of contact between different types of communities. In fact, a corollary of our discussion of hegemonic patterns is the destruction of diversity. Communities are made up of people not places. Although this may be considered heresy in geography, it seems to me that there can be a separation of political identity and territory that would negate the need for boundaries. The closest we have come to this idea in the political geography literature occurs in the debate between Pattenayak and Bayer (1987) and Laponce (1984, 1987) on territoriality and language policy. The former authors accuse Laponce of applying 'Western monism' as a universal process that is totally inapplicable to the local diversities of the three 'southern' continents. In their argument the idea of 'territorial imperative' is directly indicted as monolingual domination, that is 'politics trying to manipulate territorial boundaries to the detriment of smaller and weaker languages' (Pattenayak and Bayer, 1987: 263). Outside geography Galtung (1980: 202) has explicitly expressed an anti-territorial proposition: 'The magic of territory is waning somewhat; to defend values, styles of life . . . is becoming more meaningful than to defend geography'. This implies a contemporary process that I am not sure can be identified. In the longer term it is certainly true that while territory is necessary to facilitate pacification it has no such functional role in hot peace. But how do we get rid of territoriality?

Building hot peace

The first step in breaking out of the straitjacket of territories is to undermine the 'supernationalism' of the two power blocs. This has been the prime political goal of the European peace movement which Thompson (1987: 32) expresses very clearly:

'We need to rebuild a plural international community. . . . In Europe we might see the increasing "Finlandization" of nations in East and Central Europe and the Swedenization of nations in the West.'

This is referred to more generally as dealignment (Smith, 1987; Falk and Kalder, 1987). This is a process aimed at supplanting the blocs rather than either existing between the blocs (neutralism) or outside the blocs (nonalignment). It involves a political programme for transforming current international relations by the constitutional reassertion of state sovereignties within each bloc (Falk and Kalder, 1987: 14–15). It need not necessarily lead to withdrawal from alliance systems but the alliance systems themselves must be fundamentally altered.

Notice that this programme only attacks territorial conformity at the level of the blocs. Its tool to reach this end is a reassertion of the traditional territoriality at the level of the states. Hence dealignment is a challenge to a particular world order; it is not a direct threat to the underlying structure of the system. For Wallerstein (1984) the political Balkanization of the world is an integral necessity of the capitalist world-economy and dealignment clearly works within this geopolitical need. 'Finlandization' and 'Swedenization' do not produce Voute's cultural diversity, a Europe of neighbourhoods, but they do remove one level of pacification. To this degree it is reasonable to identify dealignment as a 'first step'.

But how do we move on to undermine the structure that inhibits peace? We must return to Galtung's (1980) basic point that the crisis of violence to which the peace movement is responding is just part of a wider crisis incorporating welfare, human rights and ecological issues. It has become a pervasive message of the European peace movement that must address these other issues. This means that the movement is moving beyond protests against armaments and alliances to become part of the global community of anti-systematic movements. Let us consider the peace movement in this light.

Anti-systematic movements are the political form that permanent resistance to the capitalist world-economy took in the nineteenth century. Initially in two branches, socialist and nationalist, the former split into communist and social democratic after 1917 to produce a legacy of three types of very successful anti-systematic movement by the mid-twentieth century. The success was measured by each type achieving a common strategic goal, the winning of state power. In the 'first world' social democratic parties and their welfare state agenda have dominated the politics; the communist movement has created its own 'second world'; and in the 'third world' anti-imperialist nationalist movements have carried all before them. But the political success has been tempered by a realization that systematic structural change seems to be no closer to being achieved. In the wake of inevitable disillusionment, new anti-systematic movements have developed and prospered in the final decades of the twentieth century. In the 'first world' a range of new movements focusing on peace, environmental and gender issues have produced different 'new lefts'. In the 'second world' human rights and pro-democracy campaigns have generated a rather uncoordinated anti-bureaucratic movement. In the 'third world' nationalist elites are being challenged by new anti-modernist and religous fundamentalist movements that reject the politics of their state leaders. These three new movements have by no means supplanted the old movements so that we now have six movements often existing as combative pairs (Wallerstein, 1988).

A common thread among the new movements is the rejection of the alienation caused by the impersonal 'bigness' inherent in old movement politics. States and parties with their remote bureaucracies have become targets. Whereas the old movements opted for a state-based strategy of political resistance the new movements are exploring alternative means of organization. In their very different ways they are challenging the territorial imperative of modern politics.

The peace movement is the least radical in this respect. Given that its prime targets are state armaments and state foreign policy, peace protest inevitably has a statecentric basis to its politics, hence the territorial-sovereignty character of its dealignment solution. Nevertheless recognition of necessary linkages to other crises enables political strategy to look beyond the states. This is most clearly seen in Galtung's (1980) 'preferred world'. He accepts the necessity of diversity of hot peace as opposed to pacification but couples this with an equity requirement. Obviously diversity within an unequal power hierarchy is very likely to generate conditions for a reversal to pacification. Hence Galtung's preferred world combines diversity and equity in what he terms a pluralist/communist social model. The communities making up the hot peace of Europe would be part of a larger global pattern in which their historic world political predominance would be gone.

At first sight the pluralist/communist social model would seem to be a simple combination of the 'best bits', or rather the ideals of, east (equity) and west (diversity). This is most certainly not the case. Galtung's arguments are of the new social movements so that top–down solutions beyond individual and community control in another variant of a 'mixed economy' is certainly not his preferred world. Bottom-up self-reliance is at the heart of this new politics. In fact this is considered to be the only way to confront successfully the alienation of the political and economic structures that impose both cold uniformity and massive inequalities. This policy imperative has implications for all peace activists including that of professionals such as geographers who wish to offer their expertise in the cause of peace.

The roles of 'peace' geographers

What are the implications of this chapter for geographers who have a concern for peace? In the introduction I noted that past handling of peace issues by political geographers has been hindered by the 'not war' conceptualization of peace. This problem is, of course, not limited to geographers. The same conundrum led Kenneth Boulding (1978) to avoid the term 'peace' in naming his original peace studies institute the 'Centre for Research on Conflict Resolution'. We have suggested a way round the problem using Braudel's social time spans to produce three alternative 'peaces'. We can apply this new framework to assess critically past roles of 'peace' geographers and suggest future roles meaningful to 'perpetual peace'. Nevertheless we will maintain the position that all three 'peaces', all three Braudellian time spans, are necessary for any comprehensive contribution to peace studies.

The peace of the *courte durée* is the work of diplomats as 'troubleshooters' as they draw lines on the map to separate protagonists. Of course, this is a very geographical activity and practical boundary-making was once possibly the most active area in political geography (Minghi, 1963). As 'experts' carrying out the policies of their masters, geographers may have had little influence in this immediate time span. In Voute's terms they are servants of the 'reptiles', employed to scar the landscape with boundaries that divide competing uniformities. It is geographers operating in other time spans who may help convert policies that produce boundaries into policies that accept borderlands as truly vital cultural frontiers. Nevertheless, even in the short term, some boundary lines will be better at preventing future violence than others. But geography boundary-makers in the past have typically been active promoters of their own states' interests. A more state-blind approach is a more peaceful approach.

The key social time span in many ways is that of the *moyenne durée*. It is here that peace studies have concentrated their efforts. From the arguments presented in this chapter, it should be apparent that we must immediately be suspicious of the phrase 'stable peace' as it appears in the titles of such works as Boulding (1978) and van der Wusten and O'Loughlin (1986). It is not just that the phrase implies pacification; it also misses the point about the challenge to peace of the Cold War geographical world order. Certainly nobody could seriously indict the Cold War for not achieving stability, at least in the 'North'. In fact, the stability is at the heart of the limitations of the 'long peace'. Such medium-term stability solutions seem to be the output of Isard's (1979/80) peace science wherein social science equilibrium system models are adapted to peace questions. Equilibrium sounds too much like the end of history. From the perspective of this, necessary 'stability' must take a back seat to 'diversity'.

This is a plea for a reversal in the traditional role of geographers in this debate. Typically, political geography solutions to world disorder have been biased towards producing larger political units implying a loss of diversity. We do not have to rely on the German geopolitik pan-region argument to sustain this point. Anti-fascist geopolitics was also prone to this error. The classic case is what DeBres (1986) describes as the 'great map scandal' of 1942 when the US geographer George Renner produced a post-war peace of enlarged states throughout the world eliminating, among others, Belgium and Switzerland. No wonder Walter Lippmann described political geography as a 'half-baked science' (DeBres, 1986: 391). A little later Griffith Taylor (1946) devised a 'geopacifics' to counteract warlike geopolitik but fell into exactly the same trap. His post-war settlement was another exercise in 'big is beautiful' with a 'power bloc' model very similar to the German pan-region solution (p. 323). Hence the dealignment model, for all its reassertions of territorial sovereignties, comes as a breath of fresh air that political geographers would do well to catch hold of. In the *moyenne durée* the contribution of political geographers can be the mapping and cherishing of diversities, our political 'non-reformist reform'.

Finally in the *longue durée* our identity as geographers dissolves into our responsibilities as world citizens. There are two arguments here. First, in circumstances of transforming structures 'experts' with their systematic knowledge are less valuable than the imagination of more informal knowledge. By definition, experience of the passing structure is not useful in the context of the new structure. Second, if we are indeed promoting a non-hierarchical new world, then experts and their hierarchy of knowledge in themselves represent a structural violence (Galtung, 1980: 415). In our *longue durée* perspective, peace professions and peace science are contradictions in terms. If change is to be from the bottom up, then political geographers must operate within the new social movements as citizens and forget the antiquated models of their time as the servants of reptiles. Moreover, this is directly relevant to how we currently teach peace issues (Jenkins, 1985).

References

Ascherson, N. (1989). Little nations hang out their flags, *The Observer*, 1 October, 1989.

Boulding, K. E. (1978). *Stable Peace*. Austin: University of Texas Press.

Braudel, F. (1980). *On History*. Chicago: University Press.

Debres, K. (1986). George Renner and the great map scandal of 1942, *Political Geography Quarterly*. **5**, 385–94.

Falk, R. and Kalder, M. (1987). Introduction. In M. Kalder and R. Falk (eds), *Dealignment*. Oxford: Blackwell, pp. 1–27.

Fukuyama, F. (1989). The end of history, *National Interest,* **16**, 3–18.

Gaddis, J. L. (1987). *The Long Peace: Inquiries into the History of the Cold War,* New York: Oxford University Press.

Gallie, W. B. (1978). *Philosophers of Peace and War*. Cambridge: Cambridge University Press.

Galtung, J. (1980). *The True Worlds. A Transnational Perspective*. New York: Free Press.

Gottmann, J. (1973). *The Significance of Territory*. Charlottesville: University Press of Virginia.

Halliday, F. (1983). *The Making of the Second Cold War*. London: Verso.

Hawkes, N. (1989). Viewpoint. *The Observer* (8 October 1989).

Isard, N. (1979/80). A definition of Peace Science, the queen of the Social Sciences (Parts I and II), *Journal of Peace Science,* **4**, 1–47 and 97–122.

Jenkins, A. (1985). Peace education and the geography curriculum. In D. Pepper and A. Jenkins (eds), *The Geography of Peace and War*. Oxford: Blackwell, pp. 202–13.

Johnston, R. J., O'Loughlin, J. and Taylor, P. J. (1987). The geography of violence and premature death: a world-systems approach. In C. Schmidt, D. Senghoas and R. Vayryned (eds), *The Quest for Peace*. London: Sage.

Langhorne, R. (1981). *The Collapse of the Concert of Europe*. London: Macmillan.

Laponce, J. A. (1984). The French language in Canada: tensions between geography and politics, *Political Geography Quarterly*, **3**, 91–104.

Laponce, J. A. (1987). More about languages and their territories: a reply to Pattanayak and Bayer, *Political Geography Quarterly,* **6**, 265–8.

Migone, G. G. (1987). The nature of bipolarity: an argument against the status quo. In M. Kalder and R. Falk (eds), *Dealignment.* Oxford: Blackwell.

Minghi, J. V. (1963). Boundary studies in political geography, *Annals, Association of American Geographers,* **53**, 407–28.

O'Tuathail, G. (1987). Beyond empiricist political geography: a comment on van der Wusten and O'Loughlin, *The Professional Geographer,* **39**, 196–7.

Pattenayak, D. P. and Bayer, J. M. (1987). Laponce's 'The French language in Canada: tensions between geography and politics' – a rejoinder. *Political Geography Quarterly,* **6**, 261–4.

Pepper, P. (1985). Geographers in search of peace. In D. Pepper and A. Jenkins (eds), *The Geography of Peace and War.* Oxford: Blackwell.

Smith, D. (1987). The Cold War. In D. Smith and E. P. Thompson (eds), *Prospectus for a Habitable Planet* London: Penguin.

Steel, R. (1980). *Walter Lippmann and the American Century.* Boston: Little, Brown.

Taylor, G. (1946). *Our Evolving Civilisation: An Introduction to Geophysics.* London: Oxford University Press.

Taylor, P. J. (1989). *Political Geography: World-Economy, Nation-State and Locality* (2nd edition). London: Longman.

Taylor, P. J. (1990). *Britain and the Cold War: 1945 as Geopolitical Transition.* London: Frances Pinter.

Thompson, E. P. (1987). The rituals of enmity. In D. Smith and E. P. Thompson (eds), *Prospectus for a Habitable Planet.* London: Penguin, pp. 11–43.

Voute, C. (1987). Whose Europe? A view from the West. In D. Smith and E. P. Thompson (eds), *Prospectus for a Habitable Planet.* London, Penguin, pp. 149–71.

Wallerstein, I. (1984). *The Politics of the World-Economy.* Cambridge: University Press.

Wallerstein, (1988). Typology of crises in the world-system. *Review,* **11**, 581–98.

Williams, C. H. (1989). The question of national congruence. In R. J. Johnston and P. J. Taylor (eds), *World in Crisis?,* Oxford: Blackwell, 196–230.

van der Wusten, H. and O'Loughlin, J. (1986). Claiming new territory for a stable peace: how geography can contribute. *The Professional Geographer,* **38**, 18–28.

van der Wusten, H. and O'Loughlin, J. (1987). Back to the future of political geography: a rejoinder to O'Tuathail. *The Professional Geographer,* **39**, 98–9.

7 Diplomatic Networks and Stable Peace

H. van der Wusten and H. van Korstanje

Introduction

Diplomats, the premises of their missions and their respective home bases are the main attributes of the oldest interstate contact network. Over time various types of people have represented their country abroad. Landed squires have acted as the plenipotentiaries of their royal rulers at a foreign court, new presidents' appointees from the business community rewarded for services rendered during election time have led embassies. The major part of the diplomatic staff positions is now manned with career diplomats educated specifically for the foreign service. Diplomatic missions have been built around elegant salons but they are now more and more situated in ordinary office blocks. The home base of the diplomatic service became the foreign office as governmental bureaucracies became distinct from the court proper, but nowadays the foreign service increasingly has also to accommodate interests from other parts of governmental bureaucracies that want to deal directly with their counterparts and kindred functional interests in other countries (military, agricultural attaches).

Diplomatic practice dates from Renaissance Italy and slowly developed as the territorial character of the state was established and a foreign policy as distinct from internal affairs became mandatory. It was codified in roughly its present form in Vienna in 1815. An update of this codification was agreed in the Vienna Convention on Diplomatic Relations in 1961. Diplomats represent their home countries. More specifically they protect the interests of their state, its lawful inhabitants and their property, they negotiate, collect information and promote friendly relations between their home country and their guest country. The establishment of a diplomatic mission implies recognition of the receiving country as a state by the sending country, but there can very well be recognition of statehood without the opening of an embassy or lower rank mission (Denza, 1976, Dembinski, 1988; Sen, 1988).

International recognition of a state is based on more or less formal criteria like the permanent control of a territory and its population by a stable, independent government. However, external recognition itself is also part of the definition of statehood. States therefore may also grant or withhold recognition on the basis of their political preferences towards the regime that asks for it. States then may only exist in the eyes of some and not in the eyes of others (Israel, China/Taiwan, East Germany, and – more extreme – South African homelands, Turkish Cyprus). Given a certain capacity and willingness of a country to engage in diplomatic relations at all, the number and rank of missions and the number and quality of diplomatic staff on the one hand reflect formal criteria of recognition and the expected flow of transactions to be handled. However, they are also the result of political decisions to stress some foreign links and not others in order to express foreign policy preferences in the international arena. Consequently they also have symbolic significance.

New means of transport and communication have changed the context of classical bilateral diplomacy. They have enabled political leaders to conduct foreign policy directly with their homologues without using their diplomatic missions. For the same reason the foreign office could tighten the links with its diplomatic missions abroad. In addition the increase in the number of intergovernmental organizations (IGOs) has opened new ways to transact diplomatic business between states. All these developments in the infrastructure of the foreign policy sector have occurred against the background of a greatly enlarged amount of international transactions facilitated by the same new technology (airplane, telecommunication). Consequently classical bilateral diplomacy has become far from centrally controlled and lost functions to direct channels between top leaders and to multilateral channels in networks that are now far more crowded.

In the second section of this chapter, we draw a rough outline of an explanatory model of a country's position in the bilateral diplomatic network taking into account previous literature (Brams, 1966; Russett and Lamb, 1969; Singer and Small, 1966; Small and Singer, 1973) and we report some features of the postwar evolution and current geographical shape of the global diplomatic network. In the third section, we discuss war/peace issues in connection with network evolution. More particularly we discuss a number of postwar conflicts, some of them degrading into actual warfare with a view to the consequences of these processes for the diplomatic network. We also look at Western Europe, one of the few regional examples of apparent stable, long-term peace in Boulding's sense with a view to the possible concomitant features of the diplomatic network. This leads to a few cautious conclusions in the fourth section. Our discussion of war/peace issues is in fact far from complete. The data base is restricted in number of cases, duration and number of points in time. In addition the quality of diplomacy, an elusive attribute, should be taken into account to draw definite conclusions about the relation of diplomacy and war/peace issues. Our data are limited to the presence and size of diplomatic missions and thus only allow us to derive some network attributes.

The chapter is based on data from a sample of ten countries. In fact it consists of five pairs that are thought to more or less represent major intuitively plausible groupings in the world. As we were not quite sure at the outset that we had categorized countries in a way useful to the problem at hand, we deliberately chose countries within each category with a number of similarities (notably size and level of development) in order to maximize chances of broad agreement between the two cases in a pair and thus maximizing chances of differences between pairs. Remaining differences within a pair could then more clearly be attributed to idiosyncrasies of the political systems of individual countries. However, in this way we of course diminished the chances of a pair to faithfully represent all the variation within a category.

We finally chose to include Belgium and the Netherlands (Western Europe); Brazil and Mexico (Latin America); Iran and Iraq (Middle East); Singapore and South Korea (Asian industrializing countries); Ethiopia and Kenya (Subsaharan Africa). We left out the major countries as they are nearly universally represented and we chose not to deal with the communist countries in this instance. Most of the time we look at the diplomatic networks constituted by these 10 countries as senders. In a few instances we also consider the reciprocal relations where our ten countries are at the receiving end.

Our data consist of registrations of presence/absence of diplomatic representations of these ten countries in most capitals of the world for 1955, 1965, 1975, 1985. For 1985 we also have an overview of diplomatic missions in terms of number of diplomats, ranks and specialists (attaches) versus general diplomats. Our set has 98 countries in 1985 and less countries within the same set at earlier dates. We particularly lack data on micro-states and small states but also on an occasional big one such as India. This has to do with the fact that the most recent data were collected from diplomatic lists kindly provided by most of the Dutch embassies abroad. In the third section, we occasionally draw upon other data derived from a more complete diplomatic dataset (collected by Nierop and Terlouw and amply analysed by Nierop (1989) and in a forthcoming book).

As diplomatic recognition presupposes mutual consent, earlier studies have tended to look at diplomatic bonds as essentially symmetric, that is reciprocal (Singer and Small, 1966, Small and Singer, 1973). In fact they often are not. Resident missions in the two countries of a pair can of course widely vary in size, but one of the missions may be non-resident or even non-existent at all and this for prolonged periods. It is by now clear that (a)symmetry is a distinctive feature of every diplomatic relation and its occurrence and pattern should be assessed. We have in this study limited our attention to resident missions and resident diplomatic personnel with official rank.

Postwar evolution and current distribution

The global diplomatic network is very partial indeed. In a matrix of 119 times 119 countries Nierop (1989) found 38 per cent of all cells with resident

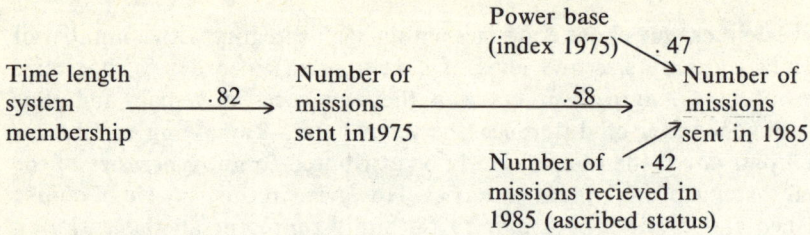

Figure 7.1   The number of diplomatic missions sent: an explanatory model

embassies, 20 per cent with non-resident, but accredited embassies and 42 per cent empty. This is not too widely different from the average situation since 1815 (Small and Singer, 1973). Currently there is a bit more emphasis on accreditation without a resident mission where the custom would have been complete negligence in former times.

Our ten countries are to a quite different degree entangled in the world diplomatic network. The Netherlands has five times more resident diplomatic missions and nine times more diplomats than Singapore. The number of missions sent is a function of various factors. It reflects earlier numbers and longevity in the international system as the building of stable international relations needs time and is largely an incremental process. There is a recurrent relation with the number of missions received reflecting the reciprocal tendency of recognition although this relation is by no means universal as we have just seen. A large representation in the outside world reflects the general power position of a country indicating a larger willingness and capacity to be engaged in the international system as power grows. Finally there is an element of deliberate policy not stemming from the preceding factors but reflecting the more or less unique position of every country in the international system that particularly differentiates between the two members of a pair. Figure 7.1 summarizes the argument and provides some support for the quality of this model by means of a series of correlations (Kendall's tau) for the various links derived from the scores of our ten countries on the variables represented in the model.

Network building in diplomacy is very much an incremental process. Overall growth in numbers of diplomatic missions 1955–85 for the ten countries was the result of 324 newly opened permanent diplomatic missions, 39 closures and 29 closures that were eventually reopened. The best predictor of the number of missions sent in 1985 is the number in 1955, our earliest recording date in this study. At that time decolonization in Asia had largely been completed but the huge increase in colonies turned into independent states had not yet occurred. The diplomatic network thus grew rapidly in the next three decades, but it kept many of its former features particularly as regards the ranking of existing states according to number of missions sent and received and the positioning of new states at the lower ranks.

There were, however, some notable exceptions to this rule among our ten countries. Mexico's number of diplomatic missions sent only grew slowly. The Mexican government was apparently less keen than the other governments in our sample countries to extend the range of its foreign contacts. This is in accordance with the results of a more general analysis (Nierop, 1989) indicating a relatively slow increase of the international links of Latin American countries. Mexico seems to be representing Latin America more faithfully in this regard than Brazil does. South Korea's diplomatic network grew very rapidly. Its insecure international status (refusal of recognition by the USSR) has led to a high profile foreign policy with the aim of establishing as many direct international relations as possible. Most of these relations in fact date from the early part of the period between 1955 and 1965.

The 1955 rank order of countries according to number of diplomatic missions sent is in its turn best seen as a reflection of the longevity of the various countries within the international system. This again stresses the incremental process of diplomatic network-building. Our sample countries in this way represent regions sequentially incorporated in the expanding European state system: Europe, Latin America, Middle East, East Asia, Africa.

Despite exceptions and delays, sending and receiving missions more often than not are reciprocal acts. In case of reciprocity countries are left with the option of having their interests represented by a non-resident mission. The relation between numbers of resident missions received and sent – in itself only a very approximate indication of reciprocity – is in fact far from perfect. Longer established, stronger countries (Netherlands, Brazil) tend to send more missions than they receive compared to more recently established weaker countries (Kenya, Ethiopia, Singapore). The major exception is again South Korea sending far more missions than it receives on account of its felt need to assure its international status. Belgium attracts many resident missions compared to the number it sends despite its general attributes because of the attraction of Brussels as the EEC centre.

As we suggested a powerful country is also an active country internationally. Differences in power within a pair frequently result in one-sided diplomatic relations with the more powerful country sending a mission. Accordingly an index of various power base elements (surface, population, GNP, military manpower) is in fact related to the number of missions sent. This is in line with the results of earlier research. Russett and Starr (1985: 199–202) have reported that size elements in particular are related to the number of diplomatic missions sent (resident and non-resident in that case).

The number of missions received indicates the ascribed status position of the country in the system. This is rather closely related to traditional roles a country has performed (e.g. United Kingdom) or additional functions it performs in the international system (e.g. Belgium). Once a country, or in fact its capital, has acquired such a status and acts as a hub in the diplomatic

Table 7.1   The distribution of diplomatic missions sent according to power position and travelling distance of the receiver in percentages per category (1985)

	Strong powers	Countries in vicinity	Distant countries
Belgium	100	90	75
Netherlands	100	87	82
Brazil	100	100	72
Mexico	100	93	44
Iran	85	81	57
Iraq	100	70	50
Singapore	100	39	4
South Korea	85	80	61
Ethiopia	85	37	26
Kenya	100	31	14

network, this may in itself be a reason for further increases of the number of missions received.

So far we have looked at the aggregates of diplomatic links countries maintain in the international system. The question then arises how countries distribute their diplomatic attention and spread their diplomatic missions geographically. As the most powerful countries send most missions, they also receive a lot of the diplomatic attention. In addition, countries tend to develop an interest in their own vicinity as many of the potential threats, opportunities and actual bonds originate from there. Rather arbitrarily we categorized seven countries as our diplomatic centres (US, Canada, UK, France, FRG, USSR, Japan). The near/distant differentiation for less powerful countries has been set at five hours' flying time. Table 7.1 shows the percentage of diplomatic missions for these various categories of countries.

There is a universal trend in the expected direction. Countries tend to be represented in all the capitals of the major countries. They then, in roughly a similar way, spread the rest of their efforts with on average a 22 points gap between the percentages filled of near and faraway countries. The main exceptions are Mexico and the Netherlands. Mexico shows a marked indifference for distant countries. This demonstrates its unusual emphasis on regional matters, very different from Brazil, the other country of this pair but probably in line with the conduct of foreign policy in many Latin American countries. The Netherlands on the other hand maintains the most widespread diplomatic network of these ten countries. Its longstanding membership of the system and long-distance colonial relations have prepared the country for worldwide involvement in international affairs even if its international role is now necessarily modest.

Missions are manned by diplomats, a handful of them in most cases. Holsti (1967: 215) suggests that today's ambassador, in contrast to his homologue of former times, 'frequently administers an embassy with a staff of several hundred specialists and secretaries'. Even American embassies are by no means generally of that size but the embassies of other nations dwarf in comparison. We called embassies large if they had ten accredited diplomats or more. Only a quarter of all embassies in this study were of that size.

Accredited diplomats are generalists of higher (ambassador, counsellor etc.) or lower (secretary) rank or specialists (attachés of various kinds). Diplomatic personnel generally are fairly evenly spread among those categories (33, 27, 40%) and this also applies roughly to most countries. The composition of embassy personnel is poorly predicted by general factors like size and development level of the receiving country and distance and exports from the perspective of the sender. There are very weak indications of high-ranking diplomats particularly residing in important countries (big, wealthy, much trade) and embassies staffed with many specialists in less important countries. Proportions of personnel sent are apparently within generally accepted but very broad rules of thumb primarily determined by highly individual attributes of countries, perhaps mainly tradition.

The size of the staff, however, is far more generally determined. In fact it seems to an extent to be guided by the same considerations as the decision to establish a mission. Table 7.2 shows strong powers to be particularly well endowed with large embassies, while nearby countries tend to have a larger proportion of these embassies than more distant countries. This last rule has some notable exceptions. For the Netherlands and South Korea the relation is inverse. The Dutch have as we saw a very widespread diplomatic network and they have a relatively larger number of big embassies outside their own region. Roughly the same applies to South Korea. A global tradition and worries about a newly acquired and insecure international status result in the same form of network. Ethiopia has only two large embassies, one in the US, reflecting the preponderance of that country in the international diplomatic network even for actual Soviet allies and the other in Italy, a strange remnant of a former imperial war and shortlived colonial experience.

Apart from the seven major countries that we arbitrarily categorized as such in Tables 7.1 and 7.2, it turned out that a few other countries were also very often the addressees of large diplomatic missions sent by our ten countries. They were Spain, Italy, China and Venezuela. This again stresses the extremely important role of Europe as the original cradle of the whole diplomatic network; it also suggests the importance of size and the specific

Table 7.2   Proportion of large size embassies in different categories of receiving countries

	Strong powers	Countries in vicinity	Distant countries
Belgium	86	17	5
Netherlands	100	12	32
Brazil	100	42	8
Mexico	100	55	20
Iran	33	8	7
Iraq	71	27	11
Singapore	14	0	0
South Korea	100	0	26
Ethiopia	17	0	7
Kenya	57	43	0

position of Venezuela as an early modernizer within Latin America.

Table 7.3 further differentiates the position of the seven major diplomatic centres as to the size of the missions sent to them and received from them in our ten sample countries. There are only two countries with serious gaps in their reciprocal relations with the major seven: South Korea for the whole period lacking recognition by the Soviet Union; and Iran, for a shorter period entangled in diplomatic disputes with the US and others. Ethiopia's lack of representation in Canada has no particular political background as far as we are aware.

The United States is clearly the foremost diplomatic centre in the world in terms of size of the missions our ten countries receive as well as in terms of the size of missions in Washington indicating the agreement between chosen role and ascribed status in the international system. The USSR's position shows more discrepancies, outranked as the country is by traditional centres UK and France and relative newcomer Japan, in terms of the size of missions sent by our sample countries. Its ascribed status seems inferior to its self-defined activist role. But it should be added that we have omitted the communist countries from our sample. This would partly redress the discrepancy.

In conclusion the diplomatic network evolves incrementally, older members of the interstate system sending the largest number of missions. The number of missions sent also reflects the degree of activism in foreign affairs that is in its turn largely dependent on the power position of a

Table 7.3   Size of diplomatic missions exchanged between ten sample countries and seven most important countries in 1985

	Canada		US		USSR		UK		Japan		France		FRG	
	S	R	S	R	S	R	S	R	S	R	S	R	S	R
Belgium	14	67	16	28	15	25	8	22	11	28	12	14	11	18
Netherlands	34	35	16	21	19	19	14	12	18	16	21	13	11	15
Brazil	64	58	16	17	37	16	11	22	25	28	24	20	16	8
Mexico	42	21	15	31	35	33	13	20	24	26	15	22	13	19
Iran			7	35	12		7	20	4	15	11	15	3	
Iraq	20	21	15	42	24	26	9	16	13	17	15	23	6	12
Singapore	11	24	?	9	6	19	5	20	4	14	3	9	4	13
South Korea	54	73			16	24	75	36	25	17	16	9	10	12
Ethiopia	8	13	13	49	9	11	4	7	5	10	4	20	6	
Kenya	17	48	9	11	23	31	8	14	10	13	10	13	9	15
Average rank	(1)	(1)	(5)	(2)	(2)	(3)	(3)	(4)	(4)	(5)	(6)	(6)	(7)	(7)

S = sent, R = received from the perspective of the sample.

country. The whole network is focused on a few strong states. They are the unavoidable addressees of missions and they also receive the largest ones. Countries tend to spread their other links and the amount of attention given to each of those links according to travelling distance.

The diplomatic network and war/peace issues

Diplomacy has of old been intimately connected with questions of war and peace. Diplomats have negotiated agreements that precluded wars, alliances to prevent or to win wars, ceasefires and peace treaties after wars had finished. The question here is different: does the structure, the form of the diplomatic network have anything to do with peace and war? And if so, as an outcome or as a condition? We start with violent or potentially violent conflict as it is easier to pinpoint and finally work our way towards possible structures of peace in the diplomatic network.

In order to find out how the diplomatic apparatus of our ten sample countries was related to the more serious conflicts in world politics, we selected a few of the longstanding disputes that have been contested during the postwar era: the international recognition of East Germany, the Israel–Arab controversy, apartheid and regional hegemony in Southern Africa, the Iran–Iraq war. These are very different conflicts as to their subject matter,

Table 7.4   Exchange of diplomatic missions with East Germany, 1955–85

| | 1955 | | 1965 | | 1975 | | 1985 | |
	S	R	S	R	S	R	S	R
Belgium					*	*	8	12
Netherlands					*	*	8	10
Brazil					*	*	2	7
Mexico					*	*	9	12
Iran					*	*	6	10
Iraq			*		*	*	5	14
Singapore								
South Korea								
Ethiopia						*	8	20
Kenya								

S = sent, R = received from the perspective of the 10 sample countries;
* = present; numbers in 1985 indicate size of mission.

and they are located differently in space and time. But they all have the actual presence or at least the hardly veiled threat of major violence (East Germany, Berlin crises 1958, 1961) about them. The exchange of diplomatic missions was the focus of the East Germany issue, non-recognition is an important point in the Israel–Arab question, the effort to isolate South Africa as a weapon to abolish apartheid has also resulted in withholding diplomatic representation, the Iran–Iraq war was initially not about the international existence of the two states, but about territory, despised regimes and religious fervour. The four problems are more or less sequential in time, although the Israel–Arab problem drags on and on.

As soon as the Western alliance had lifted the ban on East German recognition in the early 1970s, the country became part of the world diplomatic network (see Table 7.4). Only Iraq had earlier sent a mission even without receiving one in exchange. It could well be that the East Germans only opened a mission in Ethiopia after the regime of Haile Selassie had been deposed. This would have nothing to do with the earlier Western ban, but in general there has apparently been a marked willingness also outside the immediate circle of the European Western allies (Netherlands, Belgium) to comply with the requirements of American/German policy towards the recognition of East Germany and to grant recognition as soon as it was 'permitted'. The eagerness of East Germany to uphold its international status can be derived from its systematically larger missions compared to those that came her way.

Diplomatic missions exchanged with the main protagonists of the Israeli–Arab conflict as far as they have acquired statehood (Israel, Egypt, Jordan, Syria) show some traces of that conflict. West European and Latin American countries deal with both sides of the dispute. It is clearly part of the recognition perspective on the conflict that Iran and Iraq have extensive though often interrupted links with the Arab countries but not with Israel. African and Asian countries have weak connections with the parties in the Israeli–Arab conflict. On the Arab side the few missions they have maintained over the years are primarily with the most important country, Egypt. Israel has gone to a lot of trouble to prevent diplomatic isolation. In the early 1960s Israel had sent (as yet unreciprocated) missions to African countries, but they had vanished by 1975 under Arab pressure after the 1973 war. Israel has of late restored its diplomatic bonds with Kenya, one of the results of a sustained drive over the past few years to keep diplomatic contacts with African countries alive. Singapore, apparently less fearful of Arab countermeasures than South Korea, has accepted a mission from Israel and now reciprocates at chargé d'affaires level. In conclusion the diplomatic network of the Arab–Israeli conflict arena with the world at large is only very partially disturbed by the conflict. Iran and Iraq are clearly committed to one party, African countries hesitatingly chose sides, but in terms of the structure of diplomatic bonds countries in other regions largely ignore the dispute as far as possible.

More generally it is part of the diplomatic mores in the Middle East to express displeasure by downgrading a diplomatic mission or closing it altogether. In Egypt, after the change of policy towards Israel in 1977 all Arab countries and some of their allies that had missions in Cairo did this. In the mean time they have all been restored, finally those with Syria at the end of 1989. Looking over the whole period diplomatic links between Iraq and Iran on the one hand (two of our ten sample countries) and Egypt, Syria and Jordan on the other hand that had been established have very often been broken and then been reestablished. These bilateral relations have been described in Table 7.5. It turns out that in about half of the cases that there was an opportunity at all to do this, an existing link had been broken at the following recording date. This is extremely frequent compared to break-off rates in our population of countries at large (between 15 and 20 per cent for the whole period). The Israel–Arab conflict is embedded in a context in which the symbolic significance of recognition expressed in the exchange of diplomatic missions is apparently uppermost in the minds of key players in the field of foreign relations.

In Southern Africa we studied missions of our ten countries exchanged with South Africa, Angola, Mozambique and Zimbabwe. Tiny Botswana turned out only to send a mission to Belgium and to receive none from our ten countries, another indication of Brussels' privileged position in the diplomatic network. Table 7.6 shows the situation in 1985.

Western European countries keep relations on both sides of the divide, in

Table 7.5   Diplomatic missions exchanged between two sample countries and four protagonists in Israel–Arab conflict, 1955–85

	Israel		Egypt		Syria		Jordan	
	S	R	S	R	S	R	S	R
	1234	1234	1234	1234	1234	1234	1234	1234
Iran		**	***	***13	**7	**	***	
Iraq		***	***	***	**	**13	***4	

S = sent, R = received from the perspective of the 10 sample countries;
1 = 1955, 2 = 1965, 3 = 1975, 4 = 1985.

all capitals. Brazil does nearly the same but sticks to Portuguese-speaking Mozambique and Angola so far. At some point in the 1970s Brazil and South Africa both withdrew their respective missions but they have been restored since that time. Mexico does not relate to this region at all in accordance with its earlier reported emphasis on its own region. South Korea and Singapore are not represented either. The African countries and those from the Middle East shun South Africa. They concentrate their relations on Zimbabwe, the strongest black African country in the region. Black African countries so far only poorly reciprocate diplomatic missions in their direction. On 13 missions sent to them, they in turn have only sent 6. One of these from Mozambique to Ethiopia is one-sided in the opposite direction. This primarily reflects the position of Addis Ababa as OAU headquarters. Black Africans have less outward-bound diplomatic representation than Arabs in the bilateral network, but they use multilateral institutions, particularly the United Nations, for the purpose.

In conclusion, the Southern African conflict just like the Israeli–Arab dispute does not evoke uniform reactions elsewhere in the world. The regional pattern to cut diplomatic links, or not to maintain them, between the main protagonists in the region is hardly followed. Apart from African countries support for this course of action is only forthcoming from Iran and Iraq.

The Iran-Iraq war of the 1980s has provoked a number of different reactions in the diplomatic network. Apart from eventually cutting off direct bilateral diplomatic relations between the two sides, our ten countries have on the whole got more diplomatic bonds with the two countries particularly with Iraq. Growth is larger than growth between existing states in the network as a whole. Iraq has sent out new missions to the Netherlands, Belgium, Mexico and South Korea that were in the first three instances reciprocated. Iran sent out new missions to Mexico (not reciprocated) and South Korea (that had already installed a mission in Teheran). These two already rather well-connected countries extended the range of their contacts during the conflict.

On the other hand Iran's diplomatic bonds lowered in status. An

Table 7.6   Diplomatic missions exchanged between ten sample countries and four protagonists in Southern Africa, 1985

	Angola		Mozambique		Zimbabwe		South Africa	
	S	R	S	R	S	R	S	R
Belgium	2	7	1	4	4	10	3	18
Netherlands	3		6		6		8	7
Brazil	6		4				3	6
Mexico								
Iran			2		3			
Iraq					5			
Singapore								
South Korea								
Ethiopia				3	2	7		
Kenya					11	5		

S = sent, R = received from the perspective of the sample countries.

unusually high number of missions became headed by chargés d'affaires: 18 Iranian missions elsewhere, 20 foreign missions in Teheran. Only 6 of these cases were reciprocal. This suggests that motivations often differed on both sides. Diplomatic missions in Teheran tended to take a lower profile after the infringements of international rules during the occupation of the American embassy that occurred even before the war broke out. Some countries may have lowered the status of their missions to express their extreme displeasure with the infringements of diplomatic law. These missions were on average made significantly smaller than missions which remained headed by an ambassador. In 1985 Teheran had chargés d'affaires in most regions of the world, probably as the outcome of actions considered partisan or unfriendly by Iran in its actual conflict with Iraq.

The importance of diplomatic recognition and of the exchange of missions is different in the four cases. Perhaps more interesting is the difference in the uniformity of reactions. In the East German case the main supporters of East German isolation were with one exception able to impose their preference given their preponderance in the international system, and the importance of the issue in their own eyes. However, this was hardly a universal reaction as we did not include communist countries in our sample of ten countries. Such uniformity as we saw has not been demonstrated elsewhere. In the Israel–Arab conflict Arabs have only been able to pressurize the vulnerable African states, and this with limited success only. Their use of the diplomatic withdrawal weapon to force compliance within their own ranks has failed. As it is probably part of a different diplomatic culture the weapon tends not to be as sharp-edged as it is elsewhere in the first place. The Southern Africa case is similar as regards the failure to isolate South Africa. The other countries are for the moment very weakly equipped to bring their case effectively to the outside world at least in bilateral contacts. This is undoubtedly remedied to an extent by the use of multilateral forums. In the Iran–Iraq war parties were on the contrary far more concerned to extend the range of their diplomatic contacts and successfully so. Cutbacks in the status of diplomatic missions in Teheran are obviously meant to be temporary awaiting better times as a substitute for drawing even sharper lines between oneself and Iran. From the other side the intention is less clear, though current political exigencies may have inspired to make one's displeasure felt by lowering the status of one's missions rather than any long-term plan to withdraw from the diplomatic network. On the whole as seen from the partial perspective of our ten countries operating in four issues the diplomatic network seems to be only very partially orchestrated from a central point. Diplomatic moves differ somewhat between regions with respect to the scaling down and withdrawal of missions once they have been established. This probably means – although this would have to be backed up by further evidence that similar moves in different areas convey different political signals.

On the whole the pairs within a region are constrained and stimulated by the same types of general factors and they act accordingly in a roughly similar fashion. Some differences remain: Belgium attracts more diplomatic attention than the Netherlands for its central position as home of the European Community. Mexico is less concerned than Brazil over issues outside its own region. South Korea is motivated to defend and improve its international status by opening as many diplomatic channels as possible but not in trouble spots. Singapore is only interested in economically instigated diplomatic bonds and acts accordingly. As South Korea's position improves over time, differences diminish (see the position of South Korea in the Iran–Iraq war).

In some parts of the world countries are connected by an increasing number of multilateral functional agencies. This could be one of the ways in which security communities develop and a situation of stable peace based on the notion that mutually salient countries share a taboo on using violence to sort out their differences without any overarching authority is maintained. Multilateralism is thought to enhance the strength of the mutual bonds that then are more able to withstand the strains that are bound to occur in any relationship. The clearest modern example of such a development is to be found in Western Europe. The precise territorial extension of that develop-ment is however not self-evident. Is it co-existent with the EEC or the OECD countries, thus also taking in the remaining parts of Western Europe and North America and Japan?

Our interest here is in the possible influence of such a development on the

Table 7.7 Average size of diplomatic missions exchanged between Belgium and Netherlands and smaller EEC countries, other OECD members in Western Europe, other small countries in Europe, in 1985

	Small EEC		Small European OECD		Other small European	
	S	R	S	R	S	R
Belgium	6.5	7.6	6.1	6.8	5.6	11.4
Netherlands	8.5	7.1	7.7	7.6	6.8	
N =	5	5	8	8	5	5

S = sent, R = received, from the perspectives of the sample countries.

bilateral diplomatic network. We have already seen repeatedly that the seat of the EC in Brussels has clearly directed the orientation of diplomatic missions from other countries in terms of the location of resident missions and their size. The question now arises if multilateralism within the region has any influence on the bilateral channels that it overlaps. Is there a substitution effect? Table 7.7 suggests that there is probably not.

The Netherlands and Belgium, our two sample countries in the region, exchange missions with almost all countries in Europe. If we leave out the strongest countries that we also omitted in earlier tables, there is no clearcut difference in the size of missions with other EC countries, with other West European countries that are members of OECD and with other European countries (the communist ones outside OECD). All these countries are within the 5 hours' flying time range that we earlier used as a yardstick for short distances. Thus, there is no indication that bilateral diplomacy diminishes, not in terms of manpower allocated at least, as multilateralism grows. Perhaps we could say that the unusually thick tapestry of all kinds of relations woven between EC countries has found an intergovernmental corollary specifically in the outgrowth of multilateral agencies whereas an all-European bilateral diplomatic network already dating from an earlier era has been the basis for this outgrowth in the West European part of the region.

Nierop (1989) using cluster analysis as a grouping procedure for a matrix of the diplomatic links in the world as a whole has found no evidence for a specifically West European diplomatic cluster, nor for an all-European cluster for that matter. Communist countries formed one grouping, West European countries became part of a truly worldwide cluster. In conclusion there is an old all-European network that has lost some of its distinctive features but is still in place. Western Europe is now oriented worldwide in the diplomatic network and is a clearly distinctive cluster in terms of intergovernmental organizations only. This region of stable peace is built around a set of multilateral institutions based on an older bilateral very thick network that has a wider geographical extension and thus makes the zone of stable peace into an open system.

Craig (1979: 37) has stated that Europe's postwar history may be seen as a resumption of the federative policy era of the nineteenth century that was interrupted during 1866–1945. One of the shortcomings of that view is that it totally disregards the differences in geographical extension of the earlier and the more recent stages of that process. The extraordinary year 1989 might hopefully mark the beginning of an increase in the geographical extension of the current process in order to encompass all of Europe in the longer run. The bilateral diplomatic network to support that process is still in place. It might now form the basis for a new architecture of all-European multilateral agencies that then would have to support a more extended zone of stable peace than has been maintained in Western Europe after 1945.

Conclusion

The analysis stresses the traditional nature and stable features of the diplomatic network that grows incrementally and keeps its basic structure intact. At the same time it reflects and as a consequence also changes with the worldwide power configurations and with the regional context in which a country finds itself. In times of serious disputes the diplomatic system only very partially follows conflict lines in terms of links broken or missions lowered in status. In different disputes different tactics are followed in that respect, but truly worldwide cleavage lines are hardly to be found. The use of short-term diplomatic change (withdrawal etc.) as a political signal seems to be more common in some regions than in others. Its effectiveness as a signal may be questioned. Stable peace in Western Europe is apparently connected with a multitude of multilateral agencies channelling and supporting the enormous increase in mutual contacts between societies and states in that region. There is an older, very thick diplomatic network covering all Europe that may in its Western part have been instrumental in bringing about the zone of stable peace that has come into being. It may now serve as a useful base for an all-European extension of that zone.

References

Brams, S. J. (1966). Transaction flows in the international system. *American Political Science Review,* **66**, 880–98.
Craig, G. A. (1979). On the nature of diplomatic history: the relevance of some old books. In P. G. Lauren, *Diplomacy. New Approaches in History, Theory, and Policy.* New York: Free Press, Ch. 1.

Dembinski, L. (1988). *The Modern Law of Diplomacy. External Missions of States and International Organizations.* Dordrecht/Boston: Nijhoff.

Denza, E. (1976). *Diplomatic law.* New York: Oceana.

Holsti, K. J. (1967). *International Politics. A Framework for Analysis.* Prentice Hall: Englewood Cliffs N.J.

Nierop, T. (1989). International networks of diplomatic representation: some inquiries into form and trend. Paper presented at AAG, Baltimore.

Russett, B. M. and Lamb, W. C. (1969). Global patterns of diplomatic exchange 1963–1964, *Journal of Peace Research,* 37–55.

Russett, B. M. and Starr, H. (1985). *World Politics. The Menu for Choice.* San Francisco: Freeman.

Sen, B. (1988). *A Diplomat's Handbook of International Law and Practice.* Dordrecht/Boston: Nijhoff.

Singer, J. D. and Small, M. (1966). The composition and status ordering of the international system: 1815–1940, *World Politics,* **18**, 236–82.

Small, M. and Singer, J. D. (1973). Diplomatic importance of states 1816–1970: an extension and refinement of the indicator, *World Politics,* **25**, 577–99 (reprinted in J. D. Singer, *The Correlates of War* I. Free Press: New York, 1979, pp. 199–222).

Small, M. and Singer, J. D. (1982). *Resort to Arms. International and Civil Wars, 1816–1980,* Beverly Hills: Sage.

8 War, Peace and the European Community

Mark Wise

The names of the three bodies – the European Coal and Steel Community, the European Economic Community and the European Atomic Energy Community – which, since 1967, have been effectively merged into a single European Community, encouraged a tendency to see moves towards greater integration in Western Europe in narrow 'economic' terms. However, the European Community (EC) has been created by a mixture of motivations, some of which are directly concerned with issues of war and peace. Furthermore, although the history of the so-called 'Common Market' has been dominated by disputes over agricultural policy, trade barriers, budgetary contributions and similar economic matters, a concern with foreign policy and security matters now occupies an ever higher place on the EC's political agenda.

The deep roots of the 'European Idea'

For centuries, ideals of uniting the states of Europe into some kind of peaceful union have been dominated by the realities of warfare, culminating in the depraved horrors of World War II (1939–45). Nevertheless, there have always been those who continued to call for some form of union among European peoples. This 'European Idea' can be traced back to leaders such as Charlemagne who, on Christmas Day AD 800, had himself crowned Emperor of a Holy Roman Empire composed of western fragments of the once united imperial structure constructed around Rome. Although Charlemagne's edifice crumbled following his death, the ideal of European unity never died. It remained most rooted in the western continental part of Europe where the Roman cultural imprint had been most profound. Thus, in the early 1300s, the Italian Dante and the Frenchman Pierre Dubois are found espousing forms of this ideal. Centuries later the philosopher Rousseau called for a European Federation in the 1700s, while his compatriot Victor Hugo, speaking in Paris at an international Peace Conference in 1848, urged the creation of a United States of Europe on the American model. Even Britain, with its tendency to look across the oceans to its imperial possessions rather than to the continent was not completely cut off from this stream of thought. For example, in the late 1700s, William Penn, the English Quaker, wrote 'An Essay towards the Present and Future

Peace of Europe' in which he envisaged the creation of a European Parliament designed to settle interstate conflicts by peaceful means.

Of course, these were isolated voices in a world where war remained a familiar means of settling disputes and the grip of aggressive nationalism grasped ever larger masses into antagonistic groups. However, politicians of stature gradually began to promote the idea of creating European structures within which peaceful means of conflict resolution could be pursued. The slaughter in World War I (essentially a European affair) gave additional motivation to these people. Most notable was the socialist, Aristide Briand, who was French Foreign Minister for much of the 1920s. Member of a Pan-European-Union concerned to preserve the peace, he proposed, in 1929, a federal union of European states based upon a 'common market'. But the European idea enjoyed very little support, so the Great Depression of the period led inexorably into national protectionism, the rise of fascism and the disastrous descent into the abyss of World War II.

Nevertheless, the notion of unity did not perish in the maelstrom. Thus, in July 1944, a 'Draft Declaration of European Resistances' can be found stating:

> The peace of Europe is the cornerstone of the peace of the world. Within the span of a single generation Europe has been the epicentre of two world wars which have originated above all in the existence on the continent of thirty sovereign states. The anarchy can be solved only by the creation of a Federal Union among the European peoples . . . Federal Union alone can ensure the preservation of liberty and civilisation on the continent of Europe, bring about economic recovery and enable the German people to play a peaceful role in European affairs (Rossi, 1944:71–3).

It was easy to dismiss such declarations as the unrealistic aspirations of a small minority in a world torn apart by national antagonisms. But certain prominent politicians – such as Jean Monnet in France and Altiero Spinelli in Italy – were determined to promote these ideals in the post-war world. Furthermore, leaders of more nationalistic persuasions were also wedded to the notion of European unity in some guise or another. Winston Churchill's call in 1946 for 'a kind of United States of Europe' is often quoted, while the speeches of Charles de Gaulle – usually presented rather crudely as the quintessential French nationalist – contain frequent references to the need for some form of European organization in order to avoid war. In particular, he articulated the concept of a concerted 'Europe of States' in which, however, national sovereignty would be preserved rather than undermined by federalism. Despite these variations in its expression, the potency of the 'European idea' was such that it did not fade away with the end of the war, but began to mould the political map of Europe.

Clearly, the translation of this general idea into concrete political action was not achieved by one great momentous act of decision. Rather there has been an incremental 'step-by-step' process of integrating an increasing number of European states in an increasing number of ways. This development has not been driven by a common idea of what form European unity should take, nor by a single set of interests. People from across the political spectrum have supported various moves towards greater integration for diverse, sometimes contradictory motives ranging from 'moral idealism' to 'geopolitical realism', from a belief in 'free-market' economics to a desire to construct 'international socialism' across state boundaries. Action occurred when circumstances encouraged these diverse currents to flow together to form a decisive force. This happened in 1950 when the first major move towards today's Community took place.

Federalist and functionalist strategies

Before examining these steps, it is important to outline the integration theories, or political strategies, associated with these efforts to unite European states. Broadly speaking, it is possible to identify two relevant schools of thought; federalism and functionalism. Federalism has a very long intellectual history in Europe. The world wars, seen as the outcome of unbridled nationalism, stimulated a renewed interest in its concepts. In 1946 a European Union of Federalists (EUF) was formed. Essential to its strategy was the immediate creation of European federal institutions designed to weaken, even eliminate, the continent's nation-states. These institutions would lead the process of integration by directly challenging state sovereignty and stimulating a shift of loyalties from national to European level. Activists in the EUF pressed such ideas at the Hague Congress of 1948 which grouped prominent political leaders from all West-European states in an effort to promote greater unity among them. But opposition to the idea of ceding national sovereignty to a new federal structure was too great to overcome, with the result that the Congress gave birth to a very limited intergovernmental organization – the Council of Europe – rather than a United States of Europe forged by one mighty constitutional stroke.

It was a belief that such a direct federalist approach was either politically unrealistic and/or undesirable that permits us to group others in the 'functionalist' school of thought. 'Functionalism', in this context, owes its origin to David Mitrany, writing on this topic since the early 1930s (Mitrany, 1933; 1943; 1975). He was concerned with creating an international order of peace and cooperation. But he disliked the idea of creating enormous federal structures, thinking that they could easily become new superstates which would simply reproduce state rivalries on a larger 'regional' scale. He was also fearful that large federal entities would threaten individual liberties in ever more centralized societies. It was for this reason that he also rejected the idea of a confederated world government which some European federalists ultimately sought. Instead Mitrany and his followers envisaged an ever-widening network of 'functional' international agencies, each dealing with specific problems where different groups of countries could see a common interest in cooperative action. For example, states surrounding the North Sea might come together to deal with the common problems of overfishing and marine pollution, while their counterparts around the Mediterranean might set up a similar, but separate organization to cope with similar, but separate problems affecting their maritime zones. Numerous organizations could thus be set up to deal with other issues overlapping traditional state boundaries. Little by little states would find themselves interlinked in a 'spreading web of international activities and agencies' gradually eroding their effective national sovereignty and nationalistic modes of thought, making aggressive independent state action counterproductive if not impossible. Thus, the sovereign states of the world would not be challenged directly in the federalist manner – functionalists thought that this would inevitably provoke a nationalist reaction – but drawn by appeals to their self-interest into structures which allowed them to maintain the symbols of their identity but subtly undermined their independence in the cause of peace.

Step by step towards the European Community

Those who promoted the first effective steps towards the European Community of today tended to merge elements of both the federalist and functionalist thought. For example, Jean Monnet, the first major architect of Community structures, shared the federalist vision of creating a coherent United States of Europe and designed 'supranational' institutions with legislative powers designed to lead in that direction (Monnet, 1962: 203–11). However, the experience of the Hague Congress in 1948 reinforced his belief that European unity could not be achieved in rapid fashion by a few mighty constitutional strokes. Like the functionalists he sensed that unity had to be built up 'step-by-step' in ways which appealed to national self-interest and did not involve massive transfers of national sovereignty at any one time. Therefore, specific areas of activity had to be identified where states could see a mutual advantage in 'pooling' a part of their sovereignty and productive resources in a common structure governed by supranational institutions. This approach was adopted in 1950 by Robert Schuman, the French Minister for Foreign Affairs, who launched the European Coal and Steel Community (ECSC) in the following terms:

> The French government proposes to place the whole of Franco-German coal and steel production under a common High Authority, in an organisation open to the participation of other European countries. The pooling of coal and steel production will immediately ensure the establishment of a common basis for economic development – first step towards a European Federation – and change

the destiny of those regions for so long dedicated to the making of weapons for war of which they have been frequent victims. The interdependence of production thus established will make all war between France and Germany not only unthinkable, but physically impossible (Moreau 1987: 18).

This 'Schuman Plan' involved the creation of a common market within which internal national barriers to the movement of raw materials, products, people and capital related to the coal and steel industries were to be eliminated and a common external tariff imposed by the member states on imports from third countries. The interlocking of these basic industries (much more fundamental to national economies in 1950 than today) was to be managed by a distinct set of supranational institutions which revealed the larger political ambitions behind the immediate concern with coal and steel production. A High Authority, made up of people nominated by the member governments, was given the task of thinking in 'European' terms and proposing policies for the new Community. Final decision-making power, however, was to reside in a Council of Ministers made up of the relevant ministers from the member-state governments. Policy would emerge primarily out of a dialogue between the Commission and the Council, but a European Assembly made up of parliamentarians from each adhering country would play an advisory role. In addition a European Court of Justice, made up of judges nominated by all the member governments, would be the legal arbiter in the disputes arising from the operation of this political-economic Community. The laws underpinning the ECSC and its policies would of course be European in character and take precedence over any contradictory national legislation. Along with France and West Germany, Italy and the Benelux countries chose to belong to this Community, but Britain, still perceiving itself as a great world power refusing to surrender any of its political sovereignty within supranational institutions, refused.

War, peace and the ECSC

The language employed by Schuman (see above) clearly shows that the motivations were not just 'narrowly' economic in nature, but involved profound considerations of war and peace. In particular, the concern to prevent another war between France and Germany after three bloody conflicts within 70 years was obvious. Furthermore, the 'functionalist' step marked by the ECSC was unambiguously linked to an ultimate goal of creating a federated Europe within which security and prosperity could be assured for the war-torn continent. Schuman, himself a German citizen for the first 30 or so years of his life until his native border region of Lorraine was returned to France at the end of World War I, had a deep and obvious desire to reconcile France and Germany within a European entity.

But it would be naive to believe that 'moral idealism' or a narrow concern with Franco-German enmity were the sole motivations underlying this initiative. It also owed much to the wider geopolitical circumstances of the time. Following the defeat of Hitler's regime, the Allied victors were determined to ensure that German economic and military might could never again disturb the peace of Europe and the world. Hence, the enormous territorial reduction of post-war Germany and its division into occupation zones, as well as a number of more ephemeral proposals such as the 'Morgenthau Plan' (named after its American originator) which was designed to de-industrialize Germany and turn it into a peaceful pastoral country! But by the late 1940s a fear of Germany was giving way in the West to a perception of global communist threat. The progressive establishment of Communist Party control in the eastern European states occupied by the Red Army, plus dramatic East–West confrontations, like the one triggered off by the Berlin blockade of 1948, made political leaders in Washington and elsewhere keen to rebuild the shattered societies of Western Europe in order to resist the attraction of communism. Hence the introduction of 'Marshall Aid' in 1948 to channel massive US investment to Western European countries still struggling to recover from the ravages of the war. Views on the reality of a communist challenge to western security at that time vary. But, in addition to Soviet military strength, the Communist Party then held a strong electoral position in some European countries beyond the 'Iron Curtain', winning, for example, about a quarter of the popular vote in France and Italy in the first post-war decade, as well as being very influential in intellectual and trade-union circles. Clearly it was impossible to imagine a re-invigorated Western Europe capable of resisting this perceived ideological threat without the full participation of the newly formed Federal Republic of Germany composed of the American, British and French occupation zones. Memories of German military might and the appalling consequences of its deployment were still vivid among those who had endured defeat and occupation during the war. Therefore, the thought of revived German economic strength, logical as it might seem to Cold War geostrategists, raised alarm in other minds. One way of rebuilding the Federal Republic without raising the spectre of a 'Fourth Reich' emerging from the ashes of defeat was to ensure that it took place within a controlling European structure where Germans would have to manage the crucially important coal and steel sectors in cooperation with the French, Italians and others. The full significance of the Schuman Plan – the vital first step towards the European Community of today – can only be appreciated in this wider context.

A European Defence Community?

The geopolitical problem of supporting, yet containing, a West German revival in the 'interests of Western security soon extended to the more sensitive areas of defence and foreign policy. In October 1950, another French government initiative – the Pleven Plan – launched the idea of a European Defence Community (EDC) and an associated European Political Community (EPC) designed to formulate common foreign policies. Following the outbreak of the Korean War in June 1950, which brought the communist and capitalist blocs into armed confrontation, the USA wanted to rearm West Germany in order to bolster the 'European front'. The French, with memories of Nazi occupation still fresh, refused to countenance the rebuilding of a German army. Thus, with American support, they proposed the EDC as a means of reconciling conflicting demands. West Germany would rearm within a controlling set of supranational European institutions almost identical to those conceived for the ECSC. In 1952 the 'Six' states which had established the ECSC in 1951 signed treaties to set up both the EDC and the EPC. Britain once again refused to participate. However, like the USA, to which it still felt attached in a 'special relationship' between two world powers, it pressed France to ensure that these related communities were indeed established. French irritation at being directed in this rather condescending way by the 'Anglo-Saxon' powers was one reason why these schemes eventually foundered in August 1954 (Delarue, 1987, 2). Opposition from Gaullists (concerned about loss of national sovereignty) and Communists (politically committed to the Soviet Union) proved too stiff for the French government to overcome. Besides, general support for·these ambitious proposals was waning following the death of Stalin and the ending of the Korean War in 1953. Thus West German rearmament took place within the containing frameworks of the Western European Union (WEU) and, more importantly, the North Atlantic Treaty Organisation (NATO). The WEU was essentially a single-purpose 'functional' agency created in March 1948 – one year ahead of NATO – by France, Britain and the Benelux countries to ensure their collective security. It was a traditional inter-governmental body without the supranationalist elements of the ECSC, the EDC and the EPC. In 1954 it was enlarged to incorporate a rearming West Germany and Italy following the collapse of the EDC/EPC project. Over the years it became an almost moribund body completely overshadowed by the trans-Atlantic NATO edifice dominated by the USA. Thus, the vision of a supranational European body contributing to global security faded.

Economic steps towards the European Community

The architects of European unity then returned to less sensitive 'economic' matters. In 1957, the six ECSC members signed the Treaties of Rome which established the European Atomic Energy Community (EURATOM) and the much more important European Economic Community (EEC). The latter was based on a common market which was to eliminate national discriminations and introduce the free movement of goods, services, people and capital among member states. Furthermore, common economic policies were to be forged in common supranational institutions virtually identical to those set up for the ECSC. (In fact, the ECSC, the EEC and EURATOM were effectively merged in 1967 to be governed by a single set of institutions comprising the familiar pattern of a Commission to propose, a Council of Ministers to dispose, a Parliament to scrutinize and a Court of Justice to adjudicate on Community law.) Although these founding Communities were apparently restricted to the 'economic' sphere, the long-term 'political' objective of constructing a functionally interwoven Europe within which peace could be preserved had not been forgotten. The vague reference to developing 'ever closer relationships between the member states' in Article 1 of the Rome Treaty was an indication that some still shared the aim of moving towards political union.

It was this realization that strengthened British resolve to stay outside these new bodies, despite the desire of others to get them in. The potential scope of the EEC to act in a multitude of policy areas was apparent to anyone who read the Treaties carefully, while the provision for qualified majority voting in the Council of Ministers on legislation that would take precedence over national laws was unacceptable to a broad section of a British body politic determined to protect its sovereign independence. Britain tried to promote its own much looser form of European unity by setting up the European Free Trade Association (EFTA) in 1959. This organization had none of the supranational elements so strongly embedded in EC structures and was strictly concerned with free trade alone. This attracted the smaller, more peripheral states of Western Europe, but did not succeed in stopping the steady evolution of the European Community towards greater integration of ever increasing political-economic scope. In fact, less than eighteen months after setting up EFTA, the UK, at last coming to terms with its loss of world power status, applied to join the Community. After a decade of delicate diplomacy, Britain eventually joined the EC in 1973 along with Ireland and Denmark, to be followed in the 1980s by Greece (1981), Spain and Portugal (1986). At present other countries (e.g. Austria and Turkey) are seeking to join while many more (e.g. Sweden, Norway, Switzerland, Hungary, Poland) are seeking associate status of some kind).

War, peace and the European Community

What has this ever-widening scope of the Community in both political-economic and geographical terms got to do with the great issues of war and peace in Europe and the world? Put at its simplest, it can be argued that the functionalist meshing of Western European states into ever more legally interlocked activities across the whole range of governmental endeavour has made armed conflict among member states unimaginable in a way that has justified academic theorists such as Mitrany and men of action like Monnet. Disputes articulated in nationalist terms obviously still occur, inflamed by crude treatment which they receive in the mass-media. But every day, European policies are now proposed, debated, amended, decided and implemented in a routine manner which stands in staggering contrast to the situation that prevailed before the Schuman Plan set in motion the march towards the mass of multinational transactions which characterize the contemporary Community. When local government officials or business-men from as far apart as Scotland and Sardinia, Brittany and Bavaria, the Algarve and Jutland seek development assistance from common Commun-ity funds on regular trips to Brussels, or endeavour to apply EC environ-mental regulations on such things as clean water and air, they would appear to be far removed from great declarations on conflict resolution. But the drawing together of many nationalities in these prosaic common endeavours can be seen as a real contribution to peace in an historically war-torn part of the world.

Although these activities do not involve the 'high politics' of foreign policy and security, such issues have never completely disappeared from the 'European' debate. For example, President de Gaulle of France wanted loosely knit, but institutionalized European cooperation that would have security matters high on its political agenda. Thus in 1961 the so-called 'Fouchet Plan' appeared, calling for regular meetings in a permanent Council in order to harmonize foreign policies. The European Parliament already established by the EEC Treaty would be consulted, but sovereign power would remain in the hands of the Heads of State in the Council, each armed with a national veto. The plan eventually faltered in 1962, but de Gaulle never abandoned his efforts to develop such a structure. Underlying these plans was France's persistent concern to assure peace in Europe, not least by binding West Germany firmly into a European structure. This would provide a framework within which Germany's inherent strength at Europe's geographical heart could be contained and the consequences of any future reunification of its divided parts be controlled.

The European Community as a 'civilian power'?

Although de Gaulle's efforts to promote European foreign policy co-operation foundered in the face of opposition from others suspicious of French nationalism and calls for greater independence from the USA, the European Community gradually began to give expression to similar ideas. In the 1970s the concept of the EC being a 'civilian power' for peace and prosperity in world affairs enjoyed a measure of support. As Francois Duchene expressed it:

The European Community's interest as a civilian group of countries long on economic power and relatively short on armed force is as far as possible to domesticate relations between states, including those of its own members and those with states outside its frontiers. This means trying to bring to international problems the sense of responsibility and structures of contractual politics which have in the past been associated almost exclusively with 'home' and not foreign, that is, alien affairs (Duchene, 1973).

The pursuit of this 'high moral ground' in a world dominated by superpowers trying to manipulate the outcome of a multitude of regional conflicts by military means (Vietnam, Afghanistan, etc.), took place within the modest framework of European Political Cooperation (EPC) established by the EC states in 1970 (Lodge, 1989: 223–40). At first it operated in parallel with the main EC institutions, but it has gradually grown in importance to a point where it became formally incorporated into the Single European Act (SEA) of 1987 which puts it on a full Treaty basis associated with other aspects of Community activity (Council of Europe 1986). However, significant differences remain between the procedures for economic and foreign policy-making, with, for example, no provision for majority voting on the latter. The aim of EPC is to maintain a formal institutionalized process of information and consultation between member states in order to develop common action in foreign affairs; Article 30 of the SEA clearly requires that EC countries 'shall endeavour jointly to formulate and implement a European foreign policy' and strive to act 'as a cohesive force in international relations'. Although strictly national foreign policies have not been abolished, there is a commitment to consult with other member states before adopting national stances on issues of general interest. The institutional structures within which to pursue these aims are well established. At the top of the system are the regular six-monthly meetings of the European Council involving Heads of State or Government. Foreign Ministers must meet at least twice during the same period, although in practice it is more often. A Political Committee composed of senior officials manages the day-to-day business of the EPC with the help of a permanent Secretariat based in the EC Council of Ministers building in Brussels.

Conforming to the 'civilian power' concept, EPC has traditionally adopted a low-profile, with critics seeing this as a measure of ineffectiveness. Perhaps its major achievements have been within the Conference on Security and Cooperation in Europe (CSCE), notably relating to the signing of the Helsinki Final Act in 1975. This Conference, involving some 35 states from both East and West, is designed to strengthen peace in Europe by striving to enhance mutual trust and understanding. In this unwieldy context, which originally attracted little interest from the USA, the EC states formed a coherent entity able to negotiate in an effective manner with the USSR. The convening of this conference had much to do with the Soviet desire to produce a security treaty which would legitimize the post-war political map of Europe. However, the Community seized the chance offered by the CSCE to draw concessions from the Soviet Union and its allies on such things as human rights, cultural freedoms and trade across the 'Iron Curtain'. Although the atmosphere of detente in which these agreements were wrought faded in the early 1980s following the imposition of martial law in Poland, the Helsinki Final Act remained as a reference for peoples on both sides of the East–West divide to use in the struggles for human and political rights in Europe. It has played a quiet, but influential part in the dramatic collapse of single-party power in Eastern Europe which occurred in the second half of 1989.

In addition to this patient building of bridges between Eastern and Western Europe via the ECSC, the Community has also tried to play a role in settling other conflicts. For example, after an initial period of disarray following the 1973 Yom Kippur war and the resultant Arab oil embargo, the EC has developed a common policy towards the Israeli/Palestinian dispute. In notable discord with US policy, the Community adopted the 1980 Venice Declaration which, in addition to recognizing the right of all states in the region to exist within secure frontiers, clearly accepted the right of the Palestinian people to self-determination. Common EC positions, albeit often rather dilute in nature, have also been adopted with regard to the conflicts in South Africa, Central America, Afghanistan, the Falkland Isles, and Cyprus. However, their impact on the resolution of such disputes remains extremely limited.

From civilian power to military power?

This limited impact has strengthened the argument that there are limits to the exercise of 'civilian power' (Bull, 1984) and that a Community concerned with maintaining peace ultimately has to tackle the military aspects of security. First, there has been a renewed search for a more autonomous Western European security structure albeit within the overarching architecture of NATO. The reasons for this are several. Frustrations that Europeans often appear to be mere spectators of USA–USSR superpower negotiations about European security; doubts about the long-term commitment of the USA to the defence of Western Europe; worries about the instability which could develop in Eastern Europe as a result of economic failures and the political reform process; and feelings that West Europeans should be more responsible for their own security given their growing economic strength, particularly within the EC. This desire to create a 'European pillar' to counterbalance the weight of North America in the Atlantic Alliance is difficult to satisfy within the institutional framework of the EC. Problems arise because of such things as the neutrality of Ireland; the reticence of other smaller countries such as Denmark and Greece to be dragged into 'superpower politics'; Britain's attachment to a 'special relationship' with the USA, not least in the field of nuclear weapons; France's non-participation in the integrated military command structure of NATO, and its independent nuclear deterrent force; or West Germany's desire not to jeopardise movement towards German reunification. Faced with such difficulties, the moribund Western European Union (WEU) – originally established as a security organization in 1948 – was reactivated in 1984, largely in order to define a common West European position on the so-called 'Euro-missile' crisis and avoid being mere pawns in a game played by the two nuclear superpowers. The WEU provides a structure which parallels, but does not overlap exactly the political-geographical configuration of the EC. Ireland, Denmark and Greece do not belong to the WEU, while Spain and Portugal – members of the EC since 1986 – have freely chosen to adhere to it along with Britain, France, West Germany, Italy and the Benelux countries. Within the WEU, discussion of a European dimension to defence matters has become much more substantial in recent years. Its adoption of a common 'security platform' in 1987 was based on the belief that 'the construction of an integrated Europe will remain incomplete as long as it does not include security and defence' (Zanone, 1988, 10–11).

Although easy to play down the significance of the WEU, its renewed vitality should be seen as part of a larger process of strengthening economic and political integration in Western Europe mainly, but not exclusively, within EC structures. It was no accident that the reactivation of the WEU began at about the same time the first steps were being taken towards the regeneration of the Community with its aim of creating a genuine single European market by the end of 1992. In addition the Single European Act of 1987 which articulates this economic aim not only requires member states to develop a common foreign policy (see above), but also states 'that closer cooperation on questions of European security would contribute in an essential way to the development of a European identity in external policy matters'. All the signatories carry on to say that they 'are determined to maintain the technological and industrial conditions for their security' (Article 30.6), although they carefully add that 'nothing . . . shall impede closer cooperation in the field of security between certain of the (member states) within the framework of the Western European Union or the Atlantic Alliance'. Thus, the days of pretending that the EC is purely 'economic' in character and unconcerned with the 'high politics' of war and peace are over. Even neutral Ireland, not without reservations, has signed the SEA and has not been able to resist being tugged along by the larger tide of Community integration.

These moves towards more common West European security strategies have been piecemeal in character, involving different geographical clusters of countries in different organizations. Some steps have involved only two states. In the autumn of 1987 the French and West German armies cooperated in highly symbolic joint manoeuvres in Bavaria designed to underline France's commitment to the defence of the Federal Republic and promote the notion of a stronger 'European pillar' within the Atlantic

Alliance. This was followed in January 1988 by the creation of a common Franco-German Council for Defence and Security and the formation of a 4,000-strong mixed brigade under joint command (Ruhl, 1987: 11–17). President Mitterrand saw this as the potential nucleus of a common defence policy for Western Europe as a whole, while Chancellor Kohl saw it as:

> a point of crystallisation for common Franco-German security. At the end of this road there must be a common European defence with a European army including our European friends, and a trusting partnership with the United States of America (Clough 1988).

Whether the end of this particular road will ever be reached remains an open question, but even the strongly 'Atlanticist' UK government, rather dismissive of such rhetoric, has become increasingly involved in 'European' defence projects at a practical level. This is most notable in the multi-national development of weapons systems too costly for individual states to produce. Thus they cooperate, in various combinations, with the French, West Germans, Italians and Spanish in the production of military aircraft such as the Tornado, Jaguar and 'Eurofighter'. The cooperation of France, West Germany, Italy, Spain, Belgium, the Netherlands, Denmark, the UK, Ireland, Switzerland and Sweden in the European Space Agency also has a military security dimension.

A source of stability in a changing Europe?

These piecemeal moves towards a more 'European' approach to security matters have attracted limited public attention. But the dramatic changes in Eastern Europe in the latter part of 1989 (the time of writing) may well bring the relevant issues out into the open. It should be remembered that a concern with peace and stability in Europe was one of the motives behind the first steps towards today's European Community (see above). In particular, there was the French concern to bind a German renaissance inextricably into European political-economic structures; hence Robert Schuman's vision of an eventual 'European Federation' which would provide a political system capable of controlling all the potential for conflict in a part of the world containing such national and ideological diversity. Has the 'moment of truth' arrived for the 'European idea' as peoples east of the collapsing 'Iron Curtain' break free from their totalitarian systems, thus upsetting a post-war status quo based on division into two ideological blocs and a 'Cold Peace' underpinned by a nuclear balance of terror? Some, notably in France (*plus ça change, plus c'est la même chose!*) think so. As communist regimes weakened in the face of popular protest, the Frenchman Jacques Delors, President of the EC Commission, forcibly articulated the argument that movement towards full monetary and political union of the

Community should be intensified in order to provide a stable geopolitical entity around which the potentially anarchic forces unleashed in the East could be safely harnessed. In practical terms this would mean the Community providing substantial economic assistance to its 'European cousins' with a view to gradually absorbing them into some genuinely Europe-wide structure. This was a reflection of government policy in France. In his 1990 New Year address to his compatriots, the French President, François Mitterand surveyed the momentous events of 1989 and concluded:

> Either the tendency towards break-up and fragmentation will continue with the result that we will find ourselves once again in the Europe of 1919 – we know what that led to – or European unity will be built. This unity can be achieved in two stages. First, our Community of Twelve absolutely must reinforce its structures as it has just decided to do so at Strasbourg (the EC Council of December 1989) . . . The second stage still has to be invented: on the basis of the Helsinki agreements, I want to see the 1990s give birth to a European Confederation in the true sense of the term, which will associate all the states of our continent in a common, permanent organisation of exchanges, peace and security (Mitterand, 1990: 4).

It is unlikely that the present Community can be quickly expanded to absorb a new batch of Eastern European members. More possible is the creation of a European system which has a strong EC at its core, surrounded by other countries in varying degrees of associate status, rather like the EFTA states cluster around the Community at present, bound in by free-trade and other agreements. But how the security system based on a balance between the NATO and Warsaw Pact organizations will evolve remains extremely unclear; despite the ferment of ideas associated with the reform process in Eastern Europe, leaders both East and West are displaying extreme caution on these military issues.

Back to the 'German Question'

At the geopolitical heart of this debate lies 'the German Question', just as it did in the days when the EC was being launched. The French drive, supported by most of its Community partners and the USA, to intensify moves towards EC monetary and political union is motivated in part by their persistent aim of binding West Germany so tightly to its neighbours that it cannot drift off in pursuit of independent nationalist policies based on reunification, a process almost completed by summer 1990. West German politicians appeared to support this approach during the first half of 1990, seeing EC unity as the key to open the door to a unified Germany which will not be seen as a threat, economic or military, to its neighbours (Marsh, 1989). In this vision, a united Germany could become a crucial element in a 'European peace order' providing a 'social market' model of pluralistic democracy, economic prosperity and social progress at the geographical hub of Europe. Thus the 'European idea' could permit the German people to fulfil a destiny at the centre of the continent very different from the one which ended in the ruinous abyss of 1945.

It remains to be seen whether such wider visions of peaceful European unity eventually come to pass in the 1990s following the extraordinarily turbulent end to the 1980s on this long-divided continent. Those who doubt that such a political-geographical mosaic can ever be united effectively may well prove right in the end; the history of Europe gives plenty of ammunition to the pessimists. But if sceptics, weighed down by nationalist tradition, had prevailed in the early post-war years, Robert Schuman's 'first step to a European Federation' would probably not have been taken by the ECSC in 1950. Consequently, today there would probably be no European Community providing firmly established institutions through which member states can collectively cope with revolution in the Warsaw Pact countries. True, nationalist attitudes which could undermine these institutions can be found in all European countries. However, perhaps the greatest achievement of the EC is not to eradicate nationalist sentiments (an unrealistic aim in any foreseeable future) but to channel them in a way which leads more and more politicians and their publics to associate national interests with European aims. Two prominent politicians employed virtually identical formulae to express this reality. The words of West Germany's Foreign Minister, Herr Genscher, in 1989 that 'the more European our policies, the more national they will be' (Genscher, 1989) echo those of Monsieur Bosson, former French minister for EC affairs, who sought to persuade his compatriots in 1987 that 'the more one is nationalist, the more one is European' (Bosson, 1987). The preservation of peace in Europe may well depend on the degree to which such apparently paradoxical attitudes are shared by the continent's populations at large.

References

Allen, D., Rummel, R. and Wessels, W. (1982), *European Political Cooperation*, London: Butterworth.

Bosson, B. (1987), quoted in *L'Express*, 27.2.87, 59.

Bull, H. (1984), Civilian power Europe: a contradiction in terms?, *Journal of Common Market Studies*, **21** *nos* (2–3), 1982.

Clough, P. (1988), Brigade nucleus of a European Army, *The Independent*, 25.1.88.

Council of the European Communities (1986), *Single European Act and Final Act*, Brussels: Council of European Communities.

Delarue, M. (1987), 1954; Vie et mort d'une armee européenne, *Le Monde*, 24/9/87, 2.

Duchene, F. (1973), The European Community and the uncertainties of interdependence. In M. Kohnstahm and W. Hager (eds), *A Nation Writ Large?*

Foreign Policy Problems Before the European Community, London, Macmillan, pp. 1–21.

Genscher, H. D. (1989), quoted in D. Marsh (1989), Treading the German tightrope, *Financial Times*, 1.10.89.

Ginsberg, R. (1989), *Foreign Policy Actions of the European Community*, London: Adamantine.

Greenwood, D. (1988), Constructing the European pillar: issues and institutions, *Nato Review*, **3**, 13–17.

Lodge, J. (1989), European political cooperation: towards the 1990s. In J. Lodge (ed.), *The European Community and the Challenge of the Future*, London: Frances Pinter, 223–40.

Marsh, D. (1989), Treading the German tightrope, *Financial Times*, 1.10.89.

Mitrany, D. (1933), *The Progress of International Government*, London: Allen and Unwin.

Mitrany, D. (1966), *A Working Peace System, Chicago*, (the title essay was originally published in 1943), London: Royal Institute of International Affairs.

Mitrany, D. (1975), *The Functional Theory of Politics*, London: Robertson.

Mitterrand, F. (1990), *Liberation*, 1.1.90, 4.

Monnet, J. (1962), A ferment of change, *Journal of Common Market Studies*, **1**, 203–11.

Moreau, G. (1987), Discours de Robert Schuman, May 9th 1950; Paris: *La CEE*, Sirey, Paris, p. 18 (author's translation).

Rossi, E. (1944), *L'Europe de demain, a Baconnierè*, Neuchatel: La Baconnierè, pp. 71–3.

Ruhl, L. (1987), Franco-German co-operation: supportive of the Alliance and of Europe, *Nato Review*, **6**, 11–16.

Tsakaloyannis, P. (1989), The EC: from civilian power to military integration. In: J. Lodge (ed.), *The European Community and the Challenge of the Future*, London: Frances Pinter, 241–55.

Wise, M. (1989), France and European Unity. In R. Aldrich and J. Connell, *France and World Politics*, London: Routledge, pp. 35–73.

Wistrich, E. (1989), *After 1992: The United States of Europe*, Routledge, London.

Zanone, V. (1988), Europe's role in NATO, *Nato Review*, **2**, 10–11.

9 Territorial Ideology and International Conflict: The Legacy of Prior Political Formations

Alexander B. Murphy

Introduction

The clear and precise boundaries between states shown on most political maps of the world give little hint of the widespread disjunction between patterns of effective state control and the territorial aspirations of state leaders. The significance of this disjunction can be seen in the ubiquity of interstate disputes over territory. Over the past 45 years more than half of the world's states have been involved in some kind of border or territorial disagreement with another state (Day, 1987), and overlapping territorial claims are at the heart of many of the major interstate conflicts of the 1980s . . . including those between Iran and Iraq, India and Pakistan, Libya and Chad, Argentina and the United Kingdom, and Israel and its neighbours.

In a world in which state power and autonomy are tied to control over a portion of the Earth's surface (Johnston, 1986), territory is a major source of conflict between states. As Malcolm Shaw points out, 'statehood is inconceivable in the absence of a reasonably defined geographical base' (1982: 61). Territory provides the tangible framework for the exercise of power. The state area contains the resources on which the state depends and, by virtue of the generally accepted principle of state sovereignty (Gottmann 1973), delimits the portion of the Earth's surface over which the state has the international legal right to exclusive control. Control over territory is thus a basic element of a state's existence.

Interstate conflicts over territory arise because the world's more than 175 sovereign states occupy territorial units that are far from being universally accepted, untransmutable spatial givens. Instead, they are dynamic social creations resulting from competing attempts 'to affect, influence, or control people, phenomena, and relationships, by delimiting and asserting control over a geographic area' (Sack, 1986: 19). State territories have emerged out of the efforts of individuals and groups acting to achieve social and political ends through territorial control (Knight, 1982). These efforts have been driven by the desire to exercise power through the institution of the state and the recognition that state power depends upon territory. Inevitably there have been many conflicting efforts to control the same spaces and the modern political pattern is a reflection only of successful efforts at state territorial control. Behind those successes lie countless stories of failure and frustration which have given rise to territorial ideologies that are not in accord with modern political-territorial arrangements.

Territorial ideologies of this sort are held by substate minorities seeking autonomy or independence. They are also held by groups in control of the state apparatus which feel that the state should control territory outside of its present domain, and this is the central concern of this chapter. Extrastate territorial aspirations have particularly dramatic implications for the international state system (Boulding, this volume, Chapter 10). With virtually all of the land surface of the Earth already claimed by one state or another, any attempt by one state to change the territorial *status quo* through expansion necessarily threatens the territorial integrity of another state. Violence is almost always necessary to effect a change. Despite the potentially high costs, the dominant groups within states continue to press claims to territory, and many have been willing to go to war in pursuit of territorial objectives. In an effort to explain this phenomenon, several commentators have sought to identify and describe the bases of interstate territorial claims (Burghardt 1973; Hill 1976), but generalization is difficult because these bases vary so much from place to place. Claims usually arise out of some combination of economic, strategic, ethnic, historical, demographic, and physiographic considerations.

One of the most important of these is the nation-state ideal, which embodies the notion that a state's right to exercise exclusive sovereignty over an area of the Earth's surface arises from the concept of national self-determination (Knight, 1984). It is my contention that the dominance of this ideal over the past two centuries places the pursuit of territorial claims within an ideological environment in which the exercise of sovereign territorial control is legitimated in terms of historically contingent understandings of national rights. The ideological environment of legitimation, in turn, creates its own political realities (Manning and Robinson, 1985: 74) which shape the nature and development of interstate conflict over territory in fundamental ways.

In this chapter I seek to shed some light on the ideological context of interstate territorial conflict by examining the nature and significance of the relationship between historically specific understandings of the nation-state ideal and state legitimation of territorial claims. The emphasis is on the post-World War II period, but there is first a brief review of major historical developments in the relationship between the concept of national rights and territorial understandings since the rise of the modern state. Since 1945, the concept of national rights, as defined by state actors, has largely come to be divorced from the territorial aspirations of particular ethnic groups, no matter how defined. Instead, the concept of the nation is now seen as an abstraction associated with the circumstance of people living together in a politically organized area that has not been subject to territorial diminution through outside interference. Consequently, the battle of legitimate extrastate claims to territory is now primarily grounded in arguments over which state has the historical right to control the territory.

Although the idea of territorial sovereignty can be traced back to the ancient Greeks, Jean Gottmann (1973) dates the emergence of the notion of the territorial state, as presently understood, to fourteenth-century Europe. Particular communities began to think of themselves in political-territorial terms analogous to the modern conception of the state at the time, but the idea that all established large-scale political communities should have exclusive control over their territories developed only slowly over the ensuing centuries. The incorporation of this notion into international law is generally dated from the Peace of Westphalia in 1648 (Gross, 1948). The Treaties of Westphalia established the principle of state sovereignty over territory among European states that already had clearly defined boundaries (Shaw, 1982: 62). Effective control was the basis for the legitimate exercise of power over territory; despite the emergence of proto-nationalisms in such absolutist states as France, England, and Spain (Anderson, 1986: 125–7), the right to control territory in Europe was not tied to the rights of nations at that time.

The attempts of early international legal scholars, most notably Hugo Grotius (1583–1645), to elaborate principles under which one state had the right to attack the territory of another reflect the lack of concern with national rights (Hill, 1976). For Grotius, war could be justified if it was for 'defence, the obtaining of that which belongs to us or is our due, and the inflicting of punishment' (1925, 171). These were the terms on which states sought to justify territorial claims during the seventeenth century. As Norman Pounds pointed out in his review of French territorial claims over the past 400 years, France's seventeenth-century territorial claims 'were justified either on simple strategic grounds, or on grounds of inheritance and feudal right' (1954: 51). In particular, the French appealed to their own particular interpretation of the division of Charlemagne's empire in 843. The types of arguments used to justify French claims, in turn, determined the extent of the territory claimed.

The growing acceptance of natural law philosophy during the late seventeenth and early eighteenth centuries brought about a shift in the discourse of territorial claim legitimation. Under natural law, societies have the right to control the territory that is rightfully theirs according to a Divine Order, as revealed in nature and the Scriptures (Friedmann, Lissitzyn and Pugh, 1969: 4–5). In France, this idea precipitated a focus on 'les limites naturelles' of the state (Pounds, 1954: 52–3); the Pyrenees, the Alps, the Rhine, and the seas were claimed as the natural borders of the country. This argument, though not made to the exclusion of the historical claim, was a strong supplement to it. The Enlightenment introduced a new set of ideas – most notably that state sovereignty should be vested in the people. Enlightenment thinkers such as John Locke, Jean-Jacques Rousseau, and John Stuart Mill saw 'the people' as a single cultural or ethnic community sharing common political goals (van Dyke, 1977: 726–7). The people were synonymous with the nation, and each state was to derive its legitimacy from the general will of the nation (Anderson, 1986; 127). The nation-state idea, which first received concrete political expression with the French Revolution, subsequently came to dominate political thinking in Europe and elsewhere. It was important not only as a doctrine of government control by the people; it was the basis for an emerging idea that there should be a spatial congruence between nations and states.

It is hard to overestimate the importance of this aspect of the nation-state ideal, even though few states have ever come close to achieving it. If there is any common thread in the iedological fabric of modern states it is the notion that the state is, or should be, the political-territorial manifestation of the nation. Over the past 200 years, the idea that state legitimacy rests on its status as a political-territorial expression of a nation has become an unquestioned assumption of world affairs. It is reflected in the canons of international law and the rhetoric of state leaders. State actions are taken in the name of the nation even if, by any meaningful use of the term, the state does not embody a single nation.

Embedded in the nation-state ideal is the notion that state territory provides security and opportunity for the members of a nation. Territory is thus a fundamental element of nationalism itself; indeed, the age of nationalism is associated with the emergence of a deep psychological connection between nation and state territory (Gottmann, 1973). Consequently, as the nation-state idea swept Europe, it shifted the discourse of state territorial claim legitimation away from its focus on inheritance rights and natural boundaries and towards ethnic patterns. French and German claims to territory in the nineteenth century centred on arguments about national patterns: Germany based its claims on objective cultural indicators, such as language, whereas France appealed to subjective notions of common interest and self-perception (Pounds, 1954).

From the middle of the nineteenth century until 1920, the idea of the monoethnic political state reached its apogee (McNeill, 1986). The presence of different ethnic groups within European 'nation-states' at the time, if acknowledged at all, was generally thought of as a temporary phenomenon that would be swept aside in the process of nation-building. National territorial rights occupied central stage in the discourse of state territorial claim legitimation during this period, although other justifications continued to be used if they were compatible with an ethnic argument. Historical justifications for territorial control remained particularly important, because the inhabitants of territory that once belonged to a state could be argued to be part of the state's nation, actual or potential (Murphy, forthcoming).

Implicit in the nineteenth century European notion of the nation-state was the idea that only relatively large-scale societies that were organized and 'civilized' should have their own state (Knight, 1984: 171). Hence, the nation-state ideal did not stand in the way of the suppression of small nationalist movements in Europe or European colonization of much of the rest of the world. Even leaving aside small European nationalist movements and non-European societies, however, the concept of nation was hardly unproblematic. Are nations coextensive cultural communities? If so, what cultural indicators should be used? How can inherently vague cultural boundaries form the basis for distinct political boundaries? Where do people's aspirations fit in? How large or well-organized does a society have to be before it can be a nation? It took two cataclysmic wars during the first half of the twentieth century to bring the troublesome issues suggested by these questions into clearer focus.

The dominant conceptual cast to the nation-state ideal began to change immediately after 1918. The massive disruption and loss of life brought about by the struggle led to a movement away from the notion of justifiable war in international law, even if the justification was based on the nation-state ideal (Johnson, 1981: 328). The states that signed the Covenant of the League of Nations agreed 'to respect and preserve as against external aggression the territorial integrity and existing political independence of all Members of the League' (Article 10, reprinted in Friedmann, Lissitzyn, and Pugh, 1969: 917). Moreover, the decisions about the partitioning of the German, Austro-Hungarian, and Turkish empires made at the Paris Peace Conference were influenced by President Wilson's call for 'national self-determination' (Seymour, 1951). Some of Wilson's Fourteen Points could not be reconciled easily with the call for self-determination, of course, and the concept was applied inconsistently. But the Treaty of Versailles suggested a commitment to the nation-state ideal in the territorial delimitation of post-World War I Europe (see Cobban 1969: 57–84).

The complexities and uncertainties of ethnonational patterns and processes in Eastern and Central Europe made it impossible to design political units that reflected the region's nations accurately. Thus, 'from the day of their birth, these tangible embodiments of the 19th-century ideal of encasing each ethnic group within its own separate state fell painfully short of expectations' (McNeill, 1986: 61). Post-World War I Europe, which was to have experienced peace and prosperity, instead found itself confronting instability and disarray on an ever larger scale. The monoethnic nation-state ideal was illusory.

The most extreme reaction against the newly created units came from Germany, where the exaggerated nationalist ideology and aggressive policies of the Third Reich eventually led to the outbreak of World War II. The second major war of the century not only exceeded the first in terms of destruction; it highlighted the dark side of nationalism. The war brought about an even stronger rejection of the use of force in international law and it cast a deep shadow across the already tainted nineteenth century notion that a world of harmonious nation-states could be achieved simply by constructing a political pattern based on the distribution of nations defined *a priori*. The nation-state concept was not dead; it was merely changing. Over the course of the previous 150 years a world system of states had emerged from a territorial ideology in which the exercise of sovereign control over a piece of the Earth's surface was viewed as a matter of national right. That deeply embedded notion is still very much with us and fundamentally influences the legitimation of territorial claims in the post-World War II era.

The nation-state ideal and contemporary interstate territorial claims

The resilience of the nation-state concept stems from its effectiveness as an ideological foundation for the exercise of sovereign control over territory. Whatever the realities of the situation, it is almost unimaginable that those in control of the institutions of the modern state would abandon the claim that they are acting on behalf of 'the people' of the state. The idea that states are, or should be, territorial expressions of a people or nation has swept the world from its European base. Consequently, the claim of acting on behalf of the nation is universally advanced to legitimate state action, and all states are necessarily committed to the nation-state ideal, at least in an aspirational sense (Murphy, forthcoming).

Although the nation-state may be little more than a myth (Mikesell, 1983), it is unquestionably a powerful myth. The basic institutions and doctrines of international law reveal the continued vitality of the ideal (see Emerson, 1971). The United Nations, perhaps the most important institution of international law, has representatives from most of the world's states but is called the United Nations. A number of resolutions emanating from the United Nations make reference to the 'self-determination of peoples'. For example, Article 1 of the International Covenant on Civil and Political Rights specifies:

> All peoples have the right of self-determination. By virtue of that right they freely determine their political status and freely pursue their economic, social and cultural development (reprinted in Brownlie, 1967: 150).

Subsequent United Nations resolutions contain similar language (Emerson, 1971). The rhetorical commitment to self-determination in international law provides tangible evidence of the continued vitality of the Enlightenment notion of the nation-state as an abstract ideal. But the events of the first half of the twentieth century precipitated a fundamental shift in thinking at the state level about what constitutes 'the people' for whom the state theoretically exists. The shift was related to the impossibility of arriving at a generally accepted definition of a nation, the shadow that two disastrous wars of nationalism cast over ethnically based claims to territory, and the growing realization that ethnic heterogeneity was the enduring norm for most states.

The association of national sovereignty with the bloodshed of World War I and the political and social problems that beset the 'nation-states' of Europe in its aftermath brought into question the idea that the creation of political units corresponding to large-scale ethnonational patterns was a straightforward path to a better world (McNeill, 1986: 62). Correspondingly, the case for basing territorial claims on ethnonational patterns began to weaken. The rise of Nazi Germany dealt the most serious blow to the nineteenth century notion of a harmonious world of monoethnic nation-states. One of the legacies of the war, then, was the disapprobation of any extrastate claim to territory based on purely ethnic grounds.

Developments in the conduct of warfare itself have also played a role in the shift away from an ethnic basis to territorial claims. The destruction of World War II and the introduction of war technology capable of ending life on Earth combined to elevate prohibitions against aggressive armed conflict to a position of primacy in international law (Schachter, 1986). Hence, international legal statements on the right to self-determination are always qualified by injunctions against 'partial or total disruption of the national unity and territorial integrity of any other State or country' (UN Resolution 2625 (XXV) of 24 October 1970, 'Declaration of Principles of International Law Concerning Friendly Relations and Co-operation Among States in Accordance With the Charter of the United Nations', quoted in Knight, 1984: 173). In particular, the use of an ethnonational rationale for German expansionism delegitimized any association between the use of force against another state and an ethnic claim.

International injunctions against attacks on the territories of other states have not been remarkably effective, but they do make the legitimation of conflict somewhat more difficult if a state seeks to maintain at least an appearance of deferring to international law. And few states can afford to ignore international law altogether; to do so would jeopardize needed support from other states and weaken a state's efforts to encourage observance of those principles of international law that work to its advantage. Since the general principles of international law unambiguously foreclose arguments in support of extrastate territorial claims based on the distribution of ethnic or cultural groups, any state seeking to pay lip service to international law must frame the pursuit of its claim as a defensive initiative to regain territory that was wrongfully seized or appropriated.

The ethnic heterogeneity of almost all modern states presents a further obstacle to state territorial claims that are purely ethnic, as pushing an ethnic rationale to its logical conclusion could call into question the territorial integrity of most states. There is nothing new about multiethnic states, of course. But in the second half of the nineteenth century many Europeans could believe that their states were on the path toward ethnic homogeneity. It is unlikely that many monoethnic states will emerge in the foreseeable future; the creation of new states from former European colonies with complex ethnic divisions and conflicts belie a trend toward ethnic homogeneity. Moreover, the so-called ethnic revival of the 1950s and 1960s in Europe made clear that even in the hearth of the nation-state concept, nation-building is not an inevitable or unalterable historical progression.

None of this has changed the fact that state territorial ideologies are frequently in conflict with political arrangements. Consequently, states continue to press territorial claims, which they seek to legitimate in terms that will foster commitment from within and support from outside. Given the continued vitality of the nation-state ideal, the concept of national rights is still central to the legitimation of state territorial claims. But twentieth century political and social developments have precipitated a shift in the dominant view of national rights away from its prior ethnic basis. In a world with great disparities between nation and state, the concept of nation has come to be widely treated as an abstraction; internationally, nations are seen essentially as idealized social consequences of the political organization of territory.

The rise to prominence of a territorially based conception of nations in the international arena has significant implications for state legitimation of territorial claims. If nations are the social communities of states, by extension a state that loses territory to another is deprived of part of its nation. Consequently, the strongest argument that a state can raise in support of a territorial claim is that territory belonging to the state has been wrongfully taken away. This sort of historical argument has been made ever since the Peace of Westphalia. The right to restitution of that which has been taken away is widely viewed as a basic right; as Max Weber pointed out, one of the ways that humans legitimate any action is through an appeal to tradition (Connolly, 1984: 8).

The contemporary dominance of the view that nations result from the political organization of territory is reflected in the decisions and actions of international organizations. As David Knight (1984: 175) points out, despite the commitment of the United Nations to national self-determination, it has only been willing to follow through on that commitment in colonial situations. By contrast, the United Nations has consistently refused to support claims for the territorial separation of distinct national groups in Biafra, Katanga, and Kurdistan, since these territories lie within existing states. The implicit message is that Nigerians constitute a nation entitled to self-determination because of the existence of a Nigerian state, whereas the Ibo do not constitute a nation, though the Ibo fit Symmons-Symonolewicz's frequently cited definition of a nation as 'a territorially based community of human beings sharing a distinct variant of modern culture, bound together by a strong sentiment of unity and solidarity, marked by a clear historically-rooted consciousness of national identity, and possessing, or striving to possess, a genuine political self-government' (1985: 221). The disjunction between the academic view of a nation and that which dominates in world politics is obvious.

Perhaps the most outstanding indication of international acceptance of the view that political territory defines nations is the position taken by the Organization of African Unity (OAU) on African boundaries (see Touval, 1967). The OAU declared that the boundaries of African states are inviolable under the assumption that the principle of national self-determination applies only to instances of foreign domination (see Friedmann, Lissitzyn, and Pugh, 1969: 233). Thus, 'the right of the people to self-determination takes place only within the framework of the colonially defined territory' (Shaw, 1986: 191). The fact that sub-Saharan Africa has experienced fewer interstate territorial conflicts than has the Middle East, Latin America, or Asia indicates that this position has generally been accepted (van der Wusten, 1985).

The modern rules of international law on the acquisition of territory provide further evidence of the banishment of ethnic considerations as a basis for defining state territory. Aside from an agreed cession of territory from one state to another, the most important contemporary mechanism for the legal acquisition of territory by a state is prescription (see Blum, 1985). The doctrine of prescription, derived from property law, provides that title to territory can be established through continuous possession by a state convinced that the territory belongs to it, if explicit or implied acquiescence can be shown. Under this doctrine, a state's territorial rights are the consequence of historical events, not national considerations, even though the doctrine is one of the rules of a state system in which territorial sovereignty is ideologically rooted in the nation-state ideal.

As a result of the movement away from a purely ethnic view of national rights at the state level, historical arguments are now advanced in almost all contemporary interstate territorial disputes between states. In Latin America, boundary disputes are carried on amidst a flurry of competing claims about the ways in which territories were delimited during the colonial period (Child, 1985). In South Asia, arguments centre on the validity of the accession of territory to India or Pakistan following the partitioning of the subcontinent by the British (Das Gupta, 1968). In the Horn of Africa, in Southeast Asia, and along the Soviet-Chinese border, reference is made to the century-old boundaries of predecessor states or empires (Marakis, 1987; Leng, 1980; Lamb, 1968). In each case, the argument is made that the state has the right to restitution of national territory that has been wrongfully taken away.

The general prohibition in international law against the use of force against another state does not completely foreclose the historical claim; international law gives limited credence to its acceptability. International law permits self-defence against armed attack (Article 51 of the United Nations Charter, quoted in Brownlie, 1967: 16), but does not, and probably cannot, put a time limit on defensive actions (see Schachter, 1985). Hence, a state can argue that it is acting in defence of its territory when it attacks a neighbouring state in control of an area that the attacker claims was wrongfully taken at some point in history. Moreover, since international law accords explicit deference to 'the general principles of law recognized by civilized nations' (Article 38 (1) of the Statute of the International Court of Justice, quoted in Brownlie, 1967: 237), states can argue that in seeking to reacquire wrongfully taken territory, they are merely exercising a generally recognized legal right (Reisman, 1985). In support of such a claim, they can even point to instances in which territorial acquisitions made some time ago are still widely regarded as illegal. William Hough (1985) points out that this has been the position of many states with regard to the Soviet annexation of the Baltic territories in World War II.

In the international arena, then, the prior dominance of an ethnically based view of nations has given way to a territorially based conception. This change was driven by the failure of the post-World War I effort to construct a stable political arrangement based on the principle of national self-determination and the recognition that dissimilar views of ethnonational patterns and processes would lead to further global conflicts if an ethnically driven conception of national rights continued to dominate. Yet the shift has certainly not eliminated conflicts over territory; in a world in which power, autonomy, and group identity are so closely linked to the exercise of territorial sovereignty (Mann, 1984), the right to control a piece of the Earth's surface continues to be a volatile issue. Instead the shift has altered the framework for the legitimation of territorial claims with implications for the development and pursuit of extra-state claims to territory.

The contemporary role of the historical argument

Although the types of historical events that underlie interstate territorial claims vary from case to case, justifications are invariably premised on some past 'injustice' that deprived the state of territory that rightfully belongs to it. An example of the arguments that are typically advanced is a statement that appears in the introduction to an Indian atlas entitled *Chinese Aggression in Maps* (India Ministry of Information and Broadcasting, 1962). The text states:

> The first essential before we can revert to paths of peace and peaceful settlements is the undoing of all the consequences of aggression. This means that at least the *status quo* as it prevailed before the latest Chinese aggression . . . should be restored.

Claims of this sort are compatible with the now dominant understanding of nation. They derive their power from the widely accepted notion that individuals, groups, and institutions have a right to restoration of that which was wrongfully taken away. This notion, is basic in the Western concept of justice (Hart, 1961), underlies the international state system, which itself is largely a creation of the West.

That states generally legitimate territorial claims in historical terms does not mean that there are not other underlying motives. Often what lies behind the rhetoric of the legitimation of territorial claims are efforts to broaden power bases, gain control over valuable resources, provide a focus for group identity, or undermine the power of other states. Yet the necessity of framing the claim in historical terms has an impact on the development and pursuit of territorial objectives because of the powerful role of ideology and images in the creation of political realities (see Boulding, this volume, Chapter 10; Newman, this volume, Chapter 14). Most obviously, no claim, no matter how hypocritical, can be sustained in the absence of actions that lend it credibility (Schachter, 1984; 1623).

Among the most frequently cited bases for territorial claims are economic interests, strategic needs, ethnic considerations, and historical rights (Shaw, 1986; Burghardt, 1973; Hill, 1945). Each of these factors is relevant in particular cases, but the historical claim occupies a special position.

Many disputes over territory arguably have an economic basis. Ecuador claims territory in northern Peru that includes an oil-producing region, Libyan claims to a portion of northern Chad surfaced with the discovery of uranium in the area; the presence of commercially valuable phosphates in Western Sahara is almost certainly a factor in competing territorial claims there. Yet the leaders of the states involved in these disputes explicitly deny economic motives for territorial claims. Whatever legitimacy an economic claim might have had in the early twentieth century was vitiated by the association of the Japanese conquest of Manchuria and the German movement for Lebensraum with economic motives.

More often, state leaders rest their cases on the state's historical right to the territory in question. Consequently, the extent of the territory in dispute is determined by the nature of the historical claim, not by the location of the resource. For example, Colonel Khaddafi justified Libya's annexation of the Aozou strip in northern Chad in 1973 on the basis of a Franco-Italian protocol of 1935 that ostensibly allocated the strip to Libya (see Day, 1987: 114). Yet the timing of the annexation led many to conclude that Libya's real interest was in the uranium deposits in the strip.

Since economic considerations are not generally regarded as acceptable grounds for an extrastate territorial claim, a state is unlikely to pursue a claim to economically valuable land in another state unless some historical argument can be advanced in support of the claim. A map of mineral-rich border areas is likely to tell us less about the geographical pattern of boundary disputes than a historical map of the political organization of territory. Even in those rare cases in which claims to territory have been advanced with explicit reference to economic considerations, such as

Tunisia's claim in the late 1950s to a portion of Algeria (Shaw, 1986: 195–6), the claims have been abandoned.

The historical argument is also important in those instances in which strategic considerations underlie territorial claims. The territorial dispute between Japan and the Soviet Union is a case in point. One small island group (the Habomais Islands) and three other small islands (Shikotan, Kunashiri, and Etorofu) to the northeast of Hokkaido, Japan, were occupied by the Soviet Union near the end of World War II. Continued Soviet control of the sparsely inhabited islands is generally regarded as a strategic matter, but the Soviet Union bases its explicit claim to the islands on the grounds that Russia effectively controlled the islands first. The Japanese dispute this claim and further assert that they are entitled to the territory as a consequence of nineteenth and early-twentieth century treaties between Tsarist Russia and Japan (Stephan, 1974).

The significance of the discourse of claim legitimation in this case is that it limits the negotiating positions of the countries involved. The Soviet Union's strategic interests do not depend on Shikotan and the Habomais island group, since by continuing to control Kunashiri and Etorofu, the Soviets can control entrance to the Sea of Okhotsk. Yet the Soviets have conditioned return of the non-strategic islands to Japan on the signing of a formal treaty, presumably to protect itself from the claim that it is conceding the historical argument to Japan. For Japan's part, return of only some of the islands has been rejected, in part because its historical claim is based on Japanese–Russian treaties that treat the islands in question as a group. To accept only part of the group might weaken the claim to the other part.

Since the right to exercise sovereign control over territory is so closely bound up with the notion of national rights, explicit claims to territory based solely on strategic considerations are rarely made. Even the overtly strategic argument made by Israel in support of its conquest of the Golan Heights in the 1967 war was not totally divorced from historical consider-ations. Israel claimed that a modern history of consistent threat to its historic territory presented an extreme situation in which the state's very survival depended upon the control of certain surrounding territories (Drysdale and Blake, 1985: 228–9). The strategic argument was an out-growth of the historical claim; it did not stand on its own.

Contemporary interstate claims to territory on purely ethnic grounds are rare – a consequence of the delegitimation of an ethnic view of nations. An apparent exception to this is the Somali claim to control of territory in Ethiopia and Kenya on the grounds that Somali peoples inhabit these areas. Even in this unusual case (Shaw, 1986: 194–5), the historical argument is of importance. The recent feuding between Somali groups is but the latest incident in a long history of conflict among the Islamic peoples of the Horn of Africa (Prescott, 1965: 139–44). As even ardent Somali nationalists concede (Lewis, 1983), there has been little ethnic cohesion among the

Somali peoples, except during periods of intense conflict with the Amharic Christians. Not only is there no strong perceptual basis to the notion of a single Somali; there is considerable cultural diversity among the peoples of the territory that Somalia claims. The famous Russian ethnographic atlas, the Atlas Narodov Mira (Bruk and Apenchenko, 1964: plate 81), shows that the 'Somali people', as that term is used by Somali leaders, is a collection of subgroups of the Kushit group within the Semito-Hamitic family.

What becomes clear on closer examination is that the very notion of a Somali people is a conception derived from a historical argument. Somalia claims that all of 'Greater Somalia' functioned more or less as a unit during periods from the sixteenth century onward; the distinctiveness of the area seems to derive in part from the opposition of its peoples to neighbouring Amharic Christians. The movement for the creation of a 'Greater Somalia', which itself has roots in an idea advanced by the British during the colonial period, is essentially a movement for the reunification of a 'people' (see Ratcliff, 1986).

The significance of the historical basis to the ethnic claim, is that it permits the Somali state to lump together diverse peoples into a single ethnic conception. In the process, Somalia seeks to engender a greater sense of Somali unity, which has historically been achieved only when conflicts in the region have pitted the Islamic peoples as a group against the Amharic Christians. Moreover, by framing the ethnic claim in terms of reunification, the Somalis can at least operate in the grey area of the OAU's ban on interfering with the territorial integrity of post-colonial African states. Thus, the Somali experience suggests that the historical argument defines the parameters of conflict even when territorial claims are more generally associated with ethnic considerations.

Conclusion

Contemporary interstate territorial disputes take place in a world in which state rights to control territory are intimately linked to a territorially based conception of the nation-state ideal and in which warfare without provocation is widely condemned. Consequently, a state seeking to acquire territory outside of its domain cannot hope to obtain much international, or even internal, support if its articulated claim rests solely on economic, physiographic, strategic, demographic, or ethnic grounds. Instead the rallying cry for the pursuit of territorial claims is the restitution of territory that has been wrongfully taken by, or incorporated into, another state.

The historical argument is not only compatible with evolved notions of justice within the international state system; it can be a powerful force in promoting internal cohesiveness and state loyalty as well. Nationalism is backward-looking as well as forward-looking (Anderson, 1988), and the claim of a past injustice done to a national community can be an important catalyst for group loyalty to the state. Argentina's opposition to British control of the Falkland Islands, Japan's rejection of Soviet claims to islands off the coast of Hokkaido, and Guatemala's insistence on its right to control Belize have at different times acted as important rallying points for state nationalism. So important is this aspect of the ideological environment of territorial claim legitimation that a government may find it very difficult to back away from a claim once it has been made.

What this suggests is that the articulated basis for a state's territorial claim has a momentum all its own. By extension, an understanding of interstate territorial conflict cannot be premised on underlying motives alone. Words are also part of the process. The types of words chosen reflect prevailing ideas about the nature of territory and its relationship to human societies. They also have a direct bearing on the conflicts themselves. In the modern world, the simple act of framing a claim in historical terms, even if a cover for other motives, influences the parameters of acceptable conduct, the positions of the parties, and the types of solutions that are possible.

References

Anderson, J. (1986). Nationalism and geography. In J. Anderson (ed.), *The Rise of the Modern State*. Brighton: Harvester Press, pp. 115–42.

Anderson, J. (1988). Nationalist ideology and territory. In R. J. Johnston, David Knight, and Eleonore Kofman (eds.) *Nationalism, Self-determination and Political Geography*. London: Croom Helm, pp. 18–39.

Blum, Y. Z. (1965). *Historic Titles in International Law*. The Hague: Martinus Nijhoff.

Boulding, K. (1990). The nature and causes of national and military self-images in relation to war and peace. Chapter 10 of this volume.

Brownlie, I. (ed.) (1967). *Basic Documents in International Law*. Oxford: Clarendon Press.

Bruk, S. I., and Apenchenko, V. S. (eds) (1964). *Atlas Narodov Mira*. Moscow: Glavnoe Upravlenie Geodezii i Kartografii and Institut Etnografii im. N. N. Miklukno-Maklaya.

Burghardt, A. F. (1973). The bases of territorial claims. *The Geographical Review*, **63** (2), 225–45.

Child, J. (1985). *Geopolitics and Conflict in South America: Quarrels among Neighbors*. New York: Praeger.

Cobban, A. (1969). *The Nation State and National Self-Determination*. New York: Thomas Y. Crowell.

Connolly, W. (1984). Introduction: legitimacy and modernity. In William Connolly (ed.), *Legitimacy and the State*. New York: New York University Press, pp. 1–19.

Das Gupta, Jyoti Bhusan (1968). *Jammu and Kashmir*. The Hague: Martinus Nijhoff.

Day, A. J. (ed.) (1987). *Border and Territorial Disputes* (2nd edition). Harlow: Longman Group.

Diehl, P. F., and Goertz, G. (1988). Territorial changes and militarized conflict. *Journal of Conflict Resolution*, **32** (1), 103–22.

Drysdale, A. and Blake, G. H. (1985). *The Middle-East and North Africa: A Political Geography*. Oxford: Oxford University Press.

Emerson, R. (1971). Self-determination. *American Journal of International Law*, **65**, (3), 459–75.

Fitzgibbon, L. (1982). *The Betrayal of the Somalis*. London: Rex Collings.

Friedmann, W. G., Lissitzyn, O. J., and Pugh, Richard Crawford (1969). *Cases and Materials on International Law*. American casebook series. St Paul, Minn.: West Publishing Co.

Goertz, G. and Diehl, P. F. (in press). Territorial changes and recurring conflict. In Charles Gochman and Alan Sabrosky (eds), *Prisoners of War? Nation-states in the Modern Era*, Lexington, Mass.: Lexington Books.

Gottmann, J. (1973). *The Significance of Territory*. Charlottesville: University of Virginia Press.

Gross, L. (1948). The Peace of Westphalia. 1648–1948. *The American Journal of International Law*, **42** (1), 20–41.

Grotius, H. (1925). De jure belli ac pacis libritres. Volume 2. Translated by Francis W. Kelsey. In James Brown Scott (ed.), *The Classics of International Law*. Oxford: Clarendon Press.

Hart, H. L. A. (1961). *The Concept of Law*. Oxford: Clarendon Press.

Hill, N. (1976). *Claims to Territory in International Law and Relations*. Originally published in 1945 by Oxford University Press. Westport, Conn.: Greenwood Press.

Hough, W. J. H. (1985). The annexation of the Baltic states and its effect on the development of law prohibiting forcible seizure of territory. *New York Law School Journal of International and Comparative Law*, 6 (2), 300–533.

India Ministry of Information and Broadcasting (1962). *Chinese Aggression in Maps*. Delhi: Director, Publications Division, Ministry of Information and Broadcasting.

Institut National Tchadien pour les Sciences Humaines (1972). *Atlas pratiques du Tchad*. Paris: Institut Geographiques National.

Johnson, J. T. (1981). *Just War Tradition and the Restraint of War: A Moral and Historical Inquiry*. Princeton: Princeton University Press.

Johnston, R. J. (1986). Placing politics. *Political Geography Quarterly*, Supplement to **5**: S63–S78.

Knight, D. B. (1982). Identity and territory: geographical perspectives on nationalism and regionalism. *Annals of the Association of American Geographers*, **72**, 512–31.

Knight, D. B. (1984). Geographical perspectives on self-determination. In Peter Taylor and John House (eds), *Political Geography: Recent Advances and Future Directions*. London: Croom Helm, pp. 168–90.

Lamb, A. (1968). *Asian Frontiers: Studies in a Continuing Problem*. London: Pall Mall Press.

Leng, L. Y. (1980). *The Razor's Edge: Boundary Disputes in Southeast Asia*. London: Singapore University Press.

Lewis, I. M. (ed.) (1983). *Nationalism and Self-determination in the Horn of Africa*. London: Ithaca Press.

McNeill, W. H. (1986). *Poly-ethnicity and National Unity in World History*. Toronto: University of Toronto Press.

Mann, M. (1984). The autonomous power of the state: its origins, mechanisms and results. *European Journal of Sociology*, **25**, 185–213.

Manning, D. J., and Robinson, T. J. (1985). *The Place of Ideology in Political Life*. London: Croom Helm.

Marakis, J. (1987). *National and Class Conflict in the Horn of Africa*. Cambridge: Cambridge University Press.

Mikesell, M. W. (1983). The myth of the nation state. *Journal of Geography*, **82**, 257–60.

Murphy, A. B. (forthcoming). Historical justifications for territorial claims: implications for the geography of interstate conflict. *Annals of the Association of American Geographers*.

Newman, D. (1990). Overcoming the psychological barrier: the role of images in war and peace. Chapter 14 in this volume.

Pounds, N. J. G. (1954). France and les limites naturelles from the seventeenth to the 20th centuries. *Annals of the Association of American Geographers*, **44**, 51–62.

Prescott, J. R. V. (1965). *The Geography of Frontiers and Boundaries*. Chicago: Aldine.

Ratcliff, W. (1986). *Follow the Leader in the Horn: the Soviet-Cuban Presence in East Africa*. Washington, D C: The Cuban American National Foundation.

Reisman, W. M. (1985). Criteria for the lawful use of force in international law. *The Yale Journal of International Law*, **10**, (2), 279–85.

Sack, R. D. (1986). *Human Territoriality: Its Theory and History*. Cambridge: Cambridge University Press.

Schachter, O. (1984). The right of states to use armed force. *Michigan Law Review* 82 (5 and 6), 1620–46.

Schachter, O. (1985). The lawful resort to unilateral use of force. *The Yale Journal of International Law*, **10** (2), 291–4.

Schachter, O. (1986). In defense of international rules on the use of force. *The University of Chicago Law Review*, **53**, (1), 113–46.

Seymour, C. (1951). *Geography, Justice, and Politics at the Paris Conference of 1919*. New York: American Geographical Society.

Shaw, M. (1982). Territory in international law. *Netherlands Yearbook of International Law*, **13**, 61–91.

Shaw, M. (1986). *Title to Territory in Africa: International Legal Issues*. Oxford: Clarendon Press.

Stephan, J. J. (1974). *The Kuril Island: Russo-Japanese Frontier in the Pacific*. Oxford: Clarendon Press.

Symmons-Symonolewicz, K. (1985). The concept of nationhood: toward a theoretical clarification. *Canadian Review of Studies in Nationalism*, **12**, 215–22.

Touval, S. (1967). The Organization of African Unity and African borders. *International Organization*, **21**, 102–27.

van der Wusten, H. (1985). The geography of conflict since 1945. In D. Pepper and A. Jenkins, (eds), *The Geography of Peace and War*, Oxford: Basil Blackwell, pp. 13–28.

van Dyke, V. (1977). The individual, the state and ethnic communities in political theory. *World Politics*, **29**, 343–69.

10 The Nature and Causes of National and Military Self-Images in Relation to War and Peace

Kenneth E. Boulding

National states exist almost entirely in the minds of the human race in the form of images. Their frontiers are for the most part quite invisible from outer space. There are a few exceptions to this rule. A frontier with a heavy fence along it may be seen from space, and could be forested on one side and denuded on the other. There may perhaps be ill-managed grassland on one side and well managed grassland on the other. But these examples are quite rare. There are some 'natural boundaries', like rivers, such as the upper Rhine for part of its course between France and Germany, or the Rio Grande between the United States and Mexico. There are a few mountain frontiers, for instance, between France and Italy, or between Chile and Argentina, or the southern boundary of China and Tibet. Then there are coastlines which separate nations not so much from each other as from the seas and oceans, and sometimes lakes, like the Great Lakes in North America, but even these tend to be divided up by sovereignty boundaries, like the North Sea, which are quite unobservable from space.

Many boundaries are often reinforced by walls and fences which might be detectable from a very sensitive satellite – the Berlin Wall was such a barrier in the past. Some national boundaries, like that between Israel and its neighbours at present, and like those between East and West Germany in the past, do have fences that discourage migration.

Boundaries, of course, are only part of the total image of the national state, though a very important part of it. There are also images of what lies inside the boundaries in terms of a common legal system and institutions, forms of government and government organizations, languages and ethnic groups, religious structures, communication networks, patterns of economic life, and so on.

The most important of these images is the image of commitment, which is often a very important part of the image of identity of the person. The image of identity may be at least partially revealed by the answers people give to the questions, 'What are you?' or 'Who are you?' In complex societies, especially, people have several identities. One might answer to such questions, 'I am a wife, a mother, a husband, a father, a grandparent, a Presbyterian, a Rotarian, a carpenter, a manager, a government official, a tennis player, a golfer. Among these various identities, the national identity

– an American, an Englishman, a Turk, a Japanese, and so on – is usually of great importance and may even be dominant, especially in time of war. War can almost be thought of as a religious ritual of national states. Within the national state itself there may be subidentities – an Englishman rather than a Briton, a Californian or a Southerner rather than an American, a Basque rather than a Spaniard, and certainly a Kurd rather than a Turk, an Iraqi or an Iranian.

Sometimes a local identity overrides an identity with national states. As W. S. Gilbert remarked in *Utopia Limited*, 'Great Britain is that monarchy sublime, to which some add (but others do not) Ireland.' When Poland disappeared as a national state in the late eighteenth century and was partitioned between Russia, Germany, and Austria-Hungary, the Poles remained remarkably Polish and reconstituted themselves as a national state in 1919, when the three states among which they had been divided all defeated each other. One hundred and fifty years of occupation by the Turks did not make the eastern Hungarians either Turkish or Muslim, though it may have had something to do with their becoming Calvinists. Certainly in the last few hundred years the national identity has become stronger than class identities or even religious identities, although religion seems to be stronger than class in forming identities. The workers of the world do not unite and probably never will. Identity tends to be that of a particular trade, occupation, or craft rather than that of a class, perhaps because the structure of economic interests follows occupational lines much more than it does class lines. Copper miners and copper corporations have a common interest in a high relative price for copper; copper miners and aluminium miners have very little common interest.

What is puzzling here is that the national identity and commitment often cuts across economic interests. People will make enormous sacrifices for the national state with which they identify, not only in terms of taxation but also in terms of the sacrifice of life itself in war. It may be that the very geographical boundary of the national state is something like a skin. Citizens feel violated when that skin is punctured by an invading army. This can be thought of as a form of organizational rape. Zionism is an extraordinary example of the power of geographical boundaries and the intense desire for them. The Jews survived as an unbounded people, without boundaries, for 2,000 years or more.

The images in human minds can be divided roughly into 'folk' images and 'scholarly' images. Folk images are those that are learned in ordinary daily life – in the family, in the school yard, in conversations, in discussion and argument, and in the observation of the world around us. These are very important and often fairly accurate, as they can be fairly easily tested. We all have an image of the area around where we live, an image which is constantly tested by going to places and finding either that they are 'there' or 'not there'. If we go to a friend's house and find that it has burned down, then our image changes. If our friend tells us that his house has burned down and we believe him, our image changes again.

Scholarly knowledge is what we start learning in school and may also acquire partly from our parents if they are literate and educated. This is the knowledge that is transmitted mainly by formal education, also sometimes by self-education for those who are literate, who read books, go to museums and art galleries by themselves. Somewhere between folk and scholarly images are those which are provided by the communications media, such as the press, radio, and television. A distinction that somewhat overlaps the folk-scholarly distinction is that between orally transmitted and visually transmitted knowledge, including books and writings. Traditionally, folk knowledge has been mainly transmitted orally whereas scholarly knowledge is transmitted visually. Television, however, mixes the two, and gives to the nonliterate an extended oral-visual input of information which may have a profound effect on the human race, though it is hard at this juncture to know just what the effect will be.

It has been argued that television may undermine some of the more dangerous forms of political charisma. It certainly led to the decay of McCarthyism in the United States in the 1950s, and it has been argued that if we had had television at the time of Hitler he would never have come to power. The charisma which his voice exercised in large meetings and over the radio would not have carried through on television, where he would have looked rather like Charlie Chaplin, though we cannot be sure of this. Certainly the televised debates in the presidential election of 1988 in the United States suggested that television is somewhat inimical to the discussion of any important problems. There is something fundamentally domestic and therefore unscholarly about it, though, increased skills in the use of television may bring scholarship into the household on a domestic level.

Whatever the sources of images of the national state in the minds of the human race, they tend to be distortions of reality. It is not surprising that they exhibit what might be called 'perspective', the near being more vivid and more accurate than the far. Even in folk knowledge we learn that perspective is something of an illusion, that the person far down the road who looks so small to our eyes is actually about the same size as the person standing beside us whose image is so large in our eyes. Nevertheless, perspective also distorts. We may be able to see that a person far down the road is a person, but we cannot tell much about him or her. We cannot see the expression on his or her face in the way that we can the person at our side.

There is a moral perspective like visual perspective, in which the near tend to be seen as dear and the far, as not dear, or even as enemies. It is one of the prime businesses of scholarly knowledge to correct the errors of perspective, to teach us what is beyond the horizon, what the planets really look like close up, and even to interpret better what is in other people's minds. Our own mind is very near but even our spouse's mind is separated from ours by two coats of skin and an intervening ocean of air, across which we send messages by sight and sound, and maybe even touch, fragile vessels which easily get wrecked on the way. These messages are even more easily wrecked when they cross national boundaries.

Scholarly knowledge, unfortunately, does not always fulfil its mission of correcting errors of perspective. School atlases, for instance, almost universally exaggerate and pay much more attention to the country where the school lies. Surveys have shown that even in the United States half the people have only the vaguest idea of what the rest of the world is like, and even rather distorted ideas of their own country. Maps can be drawn showing the images of the world as it looks to people in different places (Gould and White, 1974). It has been said that the Crimean War (1854–56), especially in regard to Britain and Russia, was partly a result of Mercator's projection, which made Siberia look so much larger than it really is and India so much smaller. Even World War I may have had something to do with the geography books of Germany, which showed the United States on one page and Germany on one page, so people thought they were about the same size. We may not take these particular examples too seriously, but certainly the unrealistic images of world geography have something to do with the illusions that create foolish wars.

National images have many facets beyond geographical images. We have images of the national state through time, as well as images of space. The national history that is usually emphasized in schools, almost to the exclusion of the rest of the world, is depicted largely in terms of war, revolutions, and so on. Usually very little attention is paid to that 90 per cent or so of human activity which is devoted to peaceful pursuits – ploughing, sowing, reaping, making or building things, designing and inventing things, getting married, raising children, having fun, enjoying and creating drama, literature, and the arts, participating in religious exercises, communities, and so on. The nation state is presented as the centre of history as well as the centre of the world. Scholarly knowledge is perverted by the national state in so far as it governs school systems designed to turn human beings into uncomplaining conscripts willing to kill and die for a very small proportion of the human race.

Images of the national state are most likely to be prominent in the military, whose very occupation depends on the existence of the national state and also on the existence of potential or actual enemies of the national state, for without an enemy they could not justify their budgets. The development of specialized military organizations indeed, supported by forced grants, whether in kind or in money from the civilian population, goes back to Sargon and the first empires in Mesopotamia, to Egypt and the first Pharaohs. These developments are closely related to the rise of agriculture which provided the surplus of storable and transportable food beyond what the farmers needed to perpetuate themselves. An organized military threat system which could take this surplus away without destroying the maintenance of agriculture is what gave rise to empires and to national states. In early times, when the military organization did not consume the whole of the agricultural surplus, what might properly be called 'civilization', that is, the civil part of society, grew up within the shell of the military state, as it were, creating cities, roads, palaces, art, music, and eventually literature and drama. Indeed until fairly recent centuries this civil development centred around the religious organizations, which provided integrative power through persuasion and legitimation, which helped the military 'pacify' large areas, which would then develop the surplus to feed civil producers, makers of cloth, clothing, furniture, pottery, buildings, art works, books, and so on, which constitute the civil structure of civilization. Human beings will not submit to organizations for wholesale slaughter unless they feel that the slaughtering is for something that the slaughtered are against. Hence, the almost universal reliance on the part of the national state for religious legitimations to build up the national identity on which the national state and even empires depend. The inherent fragility of empires suggests that initial stages of conquest of heterogeneous peoples and territories produces a situation in which the legitimation of an empire is ultimately undermined and its military power, therefore, collapses. Of all the ancient empires, the Roman Empire was perhaps the most successful in creating Roman citizens out of heterogeneous peoples and cultures, which perhaps accounts for the fact that it lasted longer than any previous empire, even though it eventually succumbed to the erosion of legitimation which militarization always produces eventually.

The history of different parts of the world exhibits curiously different patterns in this respect. After the fall of Rome, Europe developed a mosaic of small feudal states and fragile empires, and eventually, especially after the development of the effective cannon, formed fairly stable large states, like England, Spain, France, Sweden, and Russia by the fifteenth and sixteenth centuries. Italy and Germany did not really form national states until the nineteenth century, and hence Germany especially was the battleground of larger states, with the Thirty Years' War (1618–48) as the major catastrophe.

Whether a world of culturally homogeneous national states is possible, or whether it is even desirable is a question that lies fairly far in the future. Most existing national states are multicultural, even those that have been around for a long time. The United Kingdom has the Welsh, the Scots, and the Catholics and Protestants of Northern Ireland, as well as the English. France has the Basques and the Bretons, Spain has the Basques and the Catalans; Italy has the Austrian Tyrol; and southern Italy is culturally and economically very different from northern Italy. Switzerland has four languages and at least two major religions. Even little Belgium has the Flemish and the Walloons speaking different languages. Virtually all the new African states are highly multicultural. The boundaries often divide tribes, cultures, and languages.

At the other end of the scale, Latin America, with the possible exception of Brazil and Peru, is somewhat uniform culturally and linguistically, but is divided into several states. India is not a country, but a planet, with an immense variety of languages, cultures, and religions. China is more homogeneous, but still has considerable differences between north and south, and a large number of minorities in the west, especially Tibet. The Soviet Union is a great mixture of nationalities, religions, and cultures, which not even communism can suppress. Japan is perhaps the most homogeneous national state, but even Japan has a small Korean minority. The United States and Canada are mosaics of ethnic groups; Canada is also bilingual. The existing political map of the world is clearly a result of the accidents of human history. It is astonishing that national states are so important, as they make so little sense in rational terms.

What creates and maintains a national state is often the images of its enemies. These are most likely to be prominent in the military, which depends not only on the existence of the national state, but on the existence of enemies, without which it could not justify its budget. Images in the minds of politically powerful people, such as heads of state and members of parliaments, are of exceptional importance, but these can often be influenced by charismatic leaders. It is not surprising that national states are capable of almost any kind of behaviour from the unspeakable to the benign.

Political scientists, such as Hans Morgenthau, have tried to build a theory of the international system around the concept of national interest. This indeed goes back at least to Machiavelli, who was talking for the most part about very small states, but whose ideas seem to have been copied by the large ones. National interest does not represent very much in the way of an objective reality and it is subject to constant change. The national interest indeed is what the nation is interested in, particularly from the point of view of the major decision-makers, although these people cannot usually retain power unless their images are widely shared among the population. Some states have seen their national interest as aggressive and imperialistic and go in for conquest and empire, which seems to create much more pride than it does shame in the minds even of the people who bear the burdens of such policies. Some states see their national interest as remaining neutral, staying home and minding their own business, like the modern Swedes and the Swiss. These societies are the ones that tend to get richer, as aggressiveness and imperialism tend to undermine economies by the ever-increasing absorption of the economy in an unproductive military sector. The payoffs from empire are usually not worth the costs.

Inconsistencies in the image of desired national or imperial boundaries are perhaps the most important cause of war. This may lead on the one hand to wars of conquest, as in ancient Rome or its successor, the Turkish Empire, or the European conquests of the Americas. These may pay off where what is conquered is a thinly populated culture with a lower level of productivity than the conquerors. It virtually never pays off when the conquerors and the conquered are at about the same level of development. The conqueror simply expands his boundaries to include a dissident and troublesome population, which may require constant military action, like the English conquest of Ireland. As between relatively equal powers, wars tend to represent the breakdown of the system of deterrance and have been very expensive to all the parties. World War I is a good example, as is the recent war between Iran and Iraq, which seemed to be over practically nothing but a relatively trivial boundary change.

The dynamics of the self-images of national interest are closely related to the dynamics of legitimacy, and how this changes is a very puzzling question. States which previously were aggressive and imperialistic, like Sweden until the eighteenth century, may change quite suddenly to being neutral and unaggressive. Empires can collapse almost overnight, as they have done in this century. The structure of legitimacy is not a well-behaved system, but is subject to widespread and unpredictable fluctuations. Yet this is at the core of the concept of national interest. Old legitimacies often collapse and new ones take their place. This unpredictability may be related to the existence of at least two contradictory sources of legitimacy. One is the perception of positive payoffs, or at least the hope of positive payoffs. A system in which the costs seem to exceed the benefits has a certain inherent instability, as we see, for instance, in a collapse of Catholic legitimacy in northern Europe after Luther, and in the present extraordinary decay of the legitimacy of what might be called 'fundamentalist Marxism' in the Communist countries. Oddly enough, the most stable legitimacies are often those which are hardest to test, like the promise of rewards or punishments in an afterlife. Purely secular ideologies, like Marxism, have an instability rising from the fact that they may be 'found out'. Capitalism, indeed, came very close to losing its legitimacy in the Great Depression of 1929–33, from which we recovered just in time to persuade the masses of the people that the long-run benefits of capitalism were worth legitimating.

However, there is also a contradictory source of legitimacy in the form of what I have elsewhere called the 'sacrifice trap' (Boulding, 1973). When we make sacrifices for some ideal, even for some image of the national state, it is very hard for us to admit that these have been in vain. Napoleon could still rally soldiers to his cause when he escaped from Elba, even after his enormously costly campaign in Russia. If the blood of the martyrs is the seed of the Church, the blood of the soldiers is the seed of the legitimacy of the national state, as war memorials all over the world testify. There may come a point, however, where the sacrifices become too large. Then there is revulsion, when an old legitimacy may collapse almost overnight. We saw this in the United States in the Vietnam War. It also had something to do with the collapse of both the Aztecs and the Incas, both cultures involving human sacrifice, with the Aztecs on a very large scale. Perhaps something like this happened to Sweden after the defeat by the Russians at Poltava (1709).

In the last 150 years we have seen the development of stable peace among an increasing group of independent nations, beginning perhaps in Scandinavia in the mid-nineteenth century, spreading to North America by about the 1870s, now to Western Europe, Australasia, and Japan. This involves a change in the national images of the countries involved, which requires the acceptance of existing boundaries, a minimum degree of intervention in the affairs of other nations, mutual rejection of threat as an element in the relationship, except perhaps on a very small scale.

The self-image of the military is a very important aspect of the war–peace system. Legitimated armed forces capable of acts of terrorism against civilians of another state, and occasionally against their own civilians, are what distinguish national states from all other organizations. The military organizations need to have an image of potential enemies in order to justify their existence and persuade the civilians of the state to tax themselves or suffer inflation in order to finance them, making the armed forces of the world essentially cooperative with each other ecologically, as an increase in one usually justifies an increase in a potential enemy, and so leads to arms races. Economically, therefore, they are usually competitive with their own civilians. Because of the dynamics of the system, they do tend to expand as an economy expands, until the economic pinch on the civilians is too great. Nevertheless, they are supported by the civilian population because the civilian image of the national state is also one of having potential enemies and the fear of invasion or some kind of political takeover. There may be some element of sado-masochism in all this, but it is very hard to estimate how much. It may be only marginally significant. To justify the economic sacrifices which are necessary to support the military, however, there must be among the civilians images of patriotism, self-sacrifice, and so on, which are widespread, sustained by songs, poems, speeches, rituals such as saluting the flag, veterans' days, and sacred structures like war memorials. There is sometimes, however, an implicit tension between what might be called the 'religion of the national state', in terms of patriotism and the world religions, which all emphasize the essential unity of mankind and the ultimate desirability of living in peace, even though they can very easily collaborate in wars of national states.

The question as to how national images, both of the civilian and of the military populations, can be changed to increase the expansion of the area of stable peace is crucial to the survival of the human race in this age of nuclear weapons and long-range missiles. Just as the development of the effective cannon destroyed the effectiveness and therefore the legitimacy of the feudal castle, city walls, and the institutions that went along with them, so the nuclear weapon and the long-range missile have effectively destroyed the legitimacy of national defence organizations. These organizations can no longer defend us. They can only destroy us in the long run, although they may lead to some stability in the short run. To be stable in the short run, deterrence must be unstable in the long run, for if the probability of its breaking down were zero, it would not be stable in the short run, and this would be the same thing as not having the weapons deter. Prenuclear deterrence seems to have had a probability of breakdown of something like 4 per cent per annum, rather like a 25-year flood. Nuclear deterrence may be more stable than this, but the probability of its breakdown cannot be zero or it would not deter at all. Even if the probability of breakdown is only 1 per cent per annum, like a 100-year flood, this makes a breakdown quite probable in 100 years and virtually certain in 400 years, well within historic time. When one adds to this the vulnerability of computers, or indeed any complex systems, the actual probability of nuclear breakdown may be much more than 1 per cent.

How to adjust national images, both of the general public and of the military and government, to the rude fact of a breakdown of national defence is the critical question of the next decade. This will involve a human learning process on a fairly large scale. Part of this may be accomplished by private and local enterprises – peace movements, sister city projects, personal contacts, and so on. Part of it, however, must come about through the transformation of the images of the national state in the minds of its leaders, and especially its military.

An important element in this learning process would be a revision of the image of the past in the minds of people, and especially of leaders of the present. Historians have tended to idealize war, perhaps because it is more likely to result in records, though rarely has it actually been more than 5 to 10 per cent of human activity. What is perhaps even more important has been the idealization of military victory – victors are remembered whereas the defeated are often forgotten. Nevertheless, there is a great deal of evidence that military defeat often results in both economic and cultural expansion of the defeated country, and in the relative stagnation of the victor, perhaps because of certain shifts in emphasis and legitimacy away from the military towards civilian and economic life. This principle goes back a long way. Even if we look at the earliest empires the victors went into decline and new empires came out of the defeated at the boundaries. The military success of the Turks in 1453 had something to do with the long stagnation of Islam and with the Renaissance in Europe. After the Thirty Years' War, it was Germany that became the cultural centre of Europe – in music, in philosophy, from Bach through Beethoven to Wagner, from Kant to Hegel, perhaps to Max Weber. After the defeat of Austria by Germany in 1866, Vienna experienced an extraordinary upsurge in philosophy and in all the arts. After the defeat of France by Germany in 1871, Paris became the cultural capital of the world. Music went to Paris – Debussy, Delibes, Cesar Franck; in the arts came the Impressionists, Rodin, and a great upsurge of French literature. After the great defeat of the Scots at Culloden in 1745, Edinburgh became a great cultural capital – Adam Smith, David Hume. Even Britain's loss of the American Revolution may have had something to do with the extraordinary economic upsurge of Britain after about 1780, and the great upsurge of British poetry and literature from Wordsworth perhaps through to Thomas Hardy. Counter examples, of course, can be found; perhaps Carthage. Even the total dismemberment of Poland produced Chopin. There is an important history here that still largely remains to be written.

The impact of military victory or defeat on economic development is another very interesting problem where there is a certain amount of ambiguous evidence. The bulk of the evidence suggests that military victory leads to economic as well as cultural stagnation and that defeat often has favourable economic effects. Something of a time pattern may be discernible here. In the rise of empires, the initial stage, where the imperial power consolidates a larger territory and brings relative internal peace to it, may produce a spurt of economic development. In the later stages of the empire, however, the increased military effort necessary at the frontiers weakens the overall economy and eventually leads to the collapse of the empire itself. This seems to go from Sargon right through to the British and French Empires of the nineteenth and early twentieth centuries. We could certainly argue that World War II was won economically by Japan and Germany, because their military defeat enabled them to devote a very large proportion of their available resources to getting richer, whereas the victorious powers had to sustain a large military machine which severely hampered their economic development. The Turkish Empire is certainly a good example of how military success led to economic and cultural stagnation over centuries. And while Turkey has not done spectacularly well economically since its defeat in World War I, it got rid of its empire and has probably done considerably better than in the previous centuries when it had a large empire and stagnated. The Japanese conquest of Korea delayed Japanese economic development by something like ten years. Spain is a classic example of military success that led to almost four centuries of economic stagnation. We can, of course, find some exceptions to this rule. Germany after 1871 and its victory over France continued to develop economically, even though it may have stagnated somewhat culturally. The defeat of Napoleon did not lead to much acceleration of economic development in France, partly because France's military remained active in the expansion of the French Empire overseas and diverted resources from domestic development. These interrelationships are very complex. Resources diverted from the military may not always be used wisely. There are some economic spillovers from the military, but these are apt to be very disappointing. Swords do not make very good ploughshares; they are the wrong kind of steel. The steel industry released from war has the opportunity to produce ploughshares, but may not take advantage of it. Here, as everywhere, adaptability is the secret of development. Victory tends to diminish adaptability, for there is not much need to adapt. Defeat creates an opportunity for adaptability which may not always be realized.

There are some signs of hope that we may be at the beginning of a period of great change in national images which will lead to human survival. There is an implicit recognition, even in the SDI ('Star Wars') programme in the United States and in the recent exchanges between the United States and the Soviet Union, that what might be called conventional national defence has broken down and a new system has to be found. Fortunately, the changes in national images which are necessary are neither difficult nor unprecedented. They have happened before. The very existence of stable peace and its growth around the world suggests that only a minimal number of political institutions is necessary and that war can be abolished through a change in national images rather than through the creation of a super national world government, which could easily lead to world tyranny.

There may be steps along the way here to an acceptance on the part of the military themselves of a new ethic and a new strategy. Even 'tit for tat', the solution of the prisoner's dilemma game, suggests that a military policy of extremely limited response to attack would be more successful than the unlimited response which has been characteristic of past wars. It would involve the abandonment of victory as an objective. Here a little more history of the virtues of defeat would be useful. There is also an increasing recognition among the military that aerial warfare and nuclear weapons have destroyed what might be called the traditional military ethic in terms of courage, self-sacrifice, fighting, and so on, and have turned war into the genocide of civilians and the destruction of the societies that they are supposed to defeat. The military ethic has to come to terms with this, but there seems to be no reason why it should not. And in this lies perhaps the greatest hope for the future.

The slogan of the 'New Palestinians', 'Two Nations, One Homeland', also offers great hope for the world. Large numbers of people today have seen something that Abraham, or Moses, or Jesus, or Mohammed, or Karl Marx, or Alexander, or Napoleon, or Hitler, or Stalin, or Roosevelt never saw: the view of this incredibly beautiful Earth from space. 'Two Hundred Nations, One Homeland' is perhaps the clarion call for the twenty-first century – the time of human maturity.

References

Boulding, K. E. (1973) *The Economy of Love and Fear – A Preface to Grants Economy*. Belmont, CA: Wadsworth.
Gould, P. and White, R. (1974) *Mental Maps*. New York: Penguin Books.

11 International Boundaries of Arabia: The Peaceful Resolution of Disputes

Gerald H. Blake

Introduction

International boundaries represent a sensitive interface between neighbouring states and may often be the immediate cause of international conflict, or may be used as the symbol of political tension which has little to do with the boundary line itself.

It might be anticipated that the Arabian Peninsula and its adjacent seas (the Red Sea, the Persian/Arabian Gulf and the Indian Ocean) would present an inordinate number of international boundary disputes, the primary cause of which was territorial.

First, the region has a number of indigenous tribal and territorial rivalries. Second, the boundary system on land was largely superimposed under British influence in the 1920s with scant regard for the underlying social, political and physical realities. Third, in relation to area, the region has a high concentration of international boundaries, with 8 states sharing at least 42 boundaries on land and sea. Of 13 potential land boundaries only 8 (61 per cent) have been formally agreed in whole or in part, while only 7 (24 per cent) of 29 potential maritime boundaries have been formally agreed. Fourth, there are attractive resources in key locations on land and sea including notably oil and gas, and valuable mineral deposits.

In spite of all these potentially volatile ingredients, Arabia has witnessed surprisingly few sustained armed boundary or territorial conflicts in the past 70 years. Saudi Arabia and North Yemen fought a fierce border war in 1933–34. Serious clashes occurred along the North and South Yemen boundaries in the 1920s, 1950s and 1970s, while there were tense military confrontations along the Iraq–Kuwait boundary in 1973 and 1976. Overall, however, these and other border conflicts are overshadowed by many instances of the peaceful resolution of boundary problems in Arabia.

This chapter examines five selected agreements to assess whether the territorial and functional compromises involved provide models for peaceful boundary conflict resolution elsewhere.

The Saudi Arabia-Sudan Common Zone in the Red Sea

Between 1963 and 1965, potentially commercial concentrations of metalliferous sediments such as iron, manganese, zinc, copper, silver, gold and other elements, including cadmium and cobalt, were located and identified in a number of isolated deeps in the Red Sea (Blissenbach and Nawab, 1982). By far the most promising was Atlantis II deep which lay on the Sudanese side of the median line which separates Sudan from Saudi Arabia (Figure 11.1). Metals in Atlantis II were valued at between 2.5 and 8 billion US dollars by early investigators.

An unsuccessful application to the UN was made in 1968 by a United States company for a licence to exploit the Red Sea deeps on the grounds that no state had rights there (Marston, 1984). In the same year a Saudi decree (7 September 1968) claimed Saudi ownership of all resources 'in the Red Sea bed adjacent to the Saudi continental shelf' (Lay, Churchill and Nordquist, 1973), while Sudan granted licences to a Sudanese company in partnership with Preussag AG of West Germany for further exploration.

Inter-governmental meetings began in May 1968 in an attempt to reach agreement as to Red Sea seabed sovereignty. The Saudis had hoped to involve other Red Sea riparian states, but the Khartoum Agreement of 16 May 1974 was between Sudan and Saudi Arabia only. The terms of the agreement were unique, dividing the Red Sea seabed into three zones: (a) a zone of full Saudi sovereignty extending westwards to a line where the water depth is 1000 metres; (b) a similar zone of full Sudanese sovereignty extending eastwards to a line where the water depth is 1000 metres; (c) a common zone lying below 1000 metres, including all 18 of the known metal-bearing deeps, where each state enjoys exclusive and equal rights (Figure 11.1). The parties also agreed to establish a Joint Commission to initiate exploration and exploitation of the resources of the Common Zone. Saudi Arabia agreed to fund the Commission, but would be entitled to recover such funds from royalties in due course (Article XII). Thus, the exploitation of resources which lie largely on Sudan's side of a theoretical median line was made possible by venture capital provided by Saudi Arabia.

It was a good arrangement which the parties hoped might serve as a model for other coastal states in a similar position (Mustafa, 1984). Commercial exploitation was due to begin in 1988, but the serious slump in world base metal prices made the project economically unattractive, and work was suspended in 1987.

The northern boundary of the Common Zone with Egypt could one day give rise to difficulties. While the legal Egypt–Sudan boundary (agreed in 1899) is clearly latitude 22 degrees north, the Sudanese argue that the 'administrative' boundary of 1902 should be the starting point for an offshore boundary, which would greatly favour Sudan in seabed allocation (Brownlie, 1979) (Figure 11.1). Contemporary Egyptian and Sudanese maps show their own contrasting versions of the international land boundary in

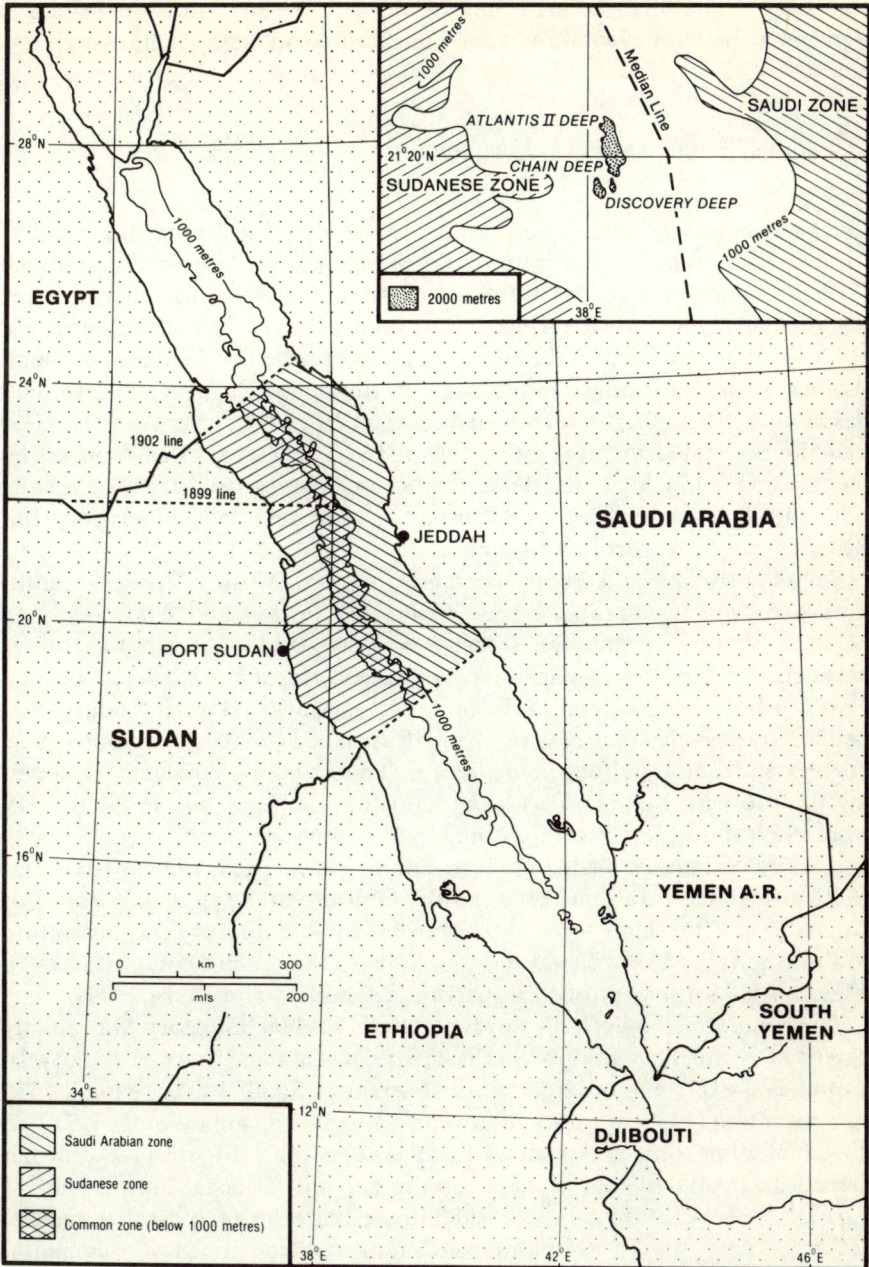

Figure 11.1 The Red Sea common zone. *Inset*: Atlantic II Deep and the median line
(After E.T. Degens and D.A. Ross, *Hot Brines and Heavy Metal Deposits in the
Red Sea*. Berlin 1969).

this sector. The offshore area may be suitable for some form of trilateral agreement between Egypt and the Joint Saudi-Sudanese Commission.

The Jordan-Saudi Arabia land boundary

During the early 1920s Britain was anxious to secure advantageous frontiers for the Kingdom of Transjordan. To the south-east lay the expansionist Wahabbi Kingdom of the Nejd under Ibn Saud, and to the south the declining Kingdom of the Hejaz. Ibn Saud declared himself King of the Hejaz in 1926 and in 1932 the two kingdoms became the Kingdom of Saudi Arabia. Even before the conquest of the Hejaz was complete, Britain had despatched Sir Gilbert Clayton to negotiate boundaries between the Nejd and the mandated territory of Transjordan. The Transjordan-Nejd settlement (The Hadda Agreement) was signed on 2 November 1925. The zig-zag boundary left the whole of Wadi Sirhan and Kaf oasis in Nejd territory, but Kaf was not to be fortified (Article 2).

Scarcely two months before the agreement, a British official, G. Antonius, had been arguing that it was 'imperative' for Kaf and Wadi Sirhan to be part of Transjordan for security reasons. Under Article 13, traders between the Nejd and Syria were guaranteed freedom of transit through Transjordan and exemption from customs duties. The boundary was defined by straight lines passing through specified points as far south as 29 degrees 35'N, on longitude 38 degrees E. There was no agreement concerning the southern boundary with the Kingdom of Hejaz but, in a letter (19 May 1927) to Ibn Saud Britain stated that the boundary should run from 29 degrees 35'N, latitude 38 degrees E to a point on the Hejaz railway two miles south of Mudauwara and thence to the Gulf of Aqaba two miles south of Aqaba town. Ibn Saud acceded to Transjordan's temporary administration of the Aqaba area, but found it 'impossible in the present circumstances to effect a final settlement of this question' (Schofield and Blake 1988).

The whole 462 mile long Jordan–Saudi Arabia boundary was finally agreed after much negotiation in the Treaty of Amman (10 August 1965). In an unusual exchange of large tracts of territory Saudi Arabia gained 4375 square miles (11331 sq km) and Jordan gained 3750 square miles (9712 sq km), including some tributary wadis of Wadi Sirhan (Figure 11.2). Jordan gained an invaluable additional 10 miles (16 km) of coast on the Gulf of Aqaba, thus facilitating the considerable expansion of coastal industrial activities near the port of Aqaba. Saudi Arabia gained a substantial geographical feature in the At Tubayq hills which had been the subject of long and detailed memoranda during the 1940s. The difficulty at that time was that the area was used for seasonal grazing by tribes from both Transjordan and Saudi Arabia. By 1965, these movements had greatly diminished from Transjordan. The Treaty of Amman also provided for the sharing of future oil revenues in a specified area (US Department of State, 1965).

Bahrain-Saudi Arabia continental shelf boundary

The Bahrain-Saudi Arabia continental shelf agreement (22 February 1958) was the first offshore agreement in the Gulf combining equidistance with equitable principles. Although a relatively short maritime boundary (98.5 nautical miles) it has some unique features which deserve attention. The southern sector of the boundary is a median line midway between pre-determined landmarks on the Bahraini and Saudi Arabian coastlines (US Department of State, 1970) (Figure 11.3). Thus, it does not follow the configuration of the coast and small islands were ignored in the delimi-tation. The middle sector was fixed in such a way that two formerly disputed islands (Lubainah al-Saghira and Lubainah al-Kabirah) were left to Bahrain and Saudi Arabia respectively. In the northern sector, the Fasht Bu Saafa hexagon was placed under Saudi sovereignty, but it was agreed (Article 2) to give Bahrain half the net income from oil revenues derived from the hexagon. The area was claimed by the Ruler of Bahrain, though on what basis is not clear. At one stage in the negotiations the principle of dividing Fasht Bu Saafa was reportedly accepted by the parties (El Hakim, 1979). Oil revenues from the Fasht Bu Saafa area have been substantial.

The Iran-Saudi Arabia continental shelf boundary

The agreement was concluded on 24 October 1968, delimiting the longest (139 nautical miles) maritime boundary in the Gulf. The agreement is of great significance in terms of the development of the law of continental shelf delimitation. According to S.H. Amin, it 'can serve as a model indicating how to divide amicably the submarine areas in narrow seas between opposite states' (Amin, 1981). The two states had experienced genuine difficulties over their offshore boundary since 1948. Concessions granted to oil companies overlapped. Ownership of two small uninhabited islands at the centre of the Gulf was in dispute. Moreover, Iran refused to ratify a draft agreement initialled in December 1965 after fresh discoveries of oil in the northern sector of the proposed delimitations. Nearly three years of further negotiations were necessary before a revised agreement was signed in October 1968. During this time, attempts were made to estimate the volume of oil in the disrupted northern sector. An accurate coastal map of the two states was also prepared by United States surveyors.

The 1968 boundary line was basically determined according to equi-distance principles, with important modifications to achieve an equitable division of oil resources (Figure 11.4). Farsi island was recognized as

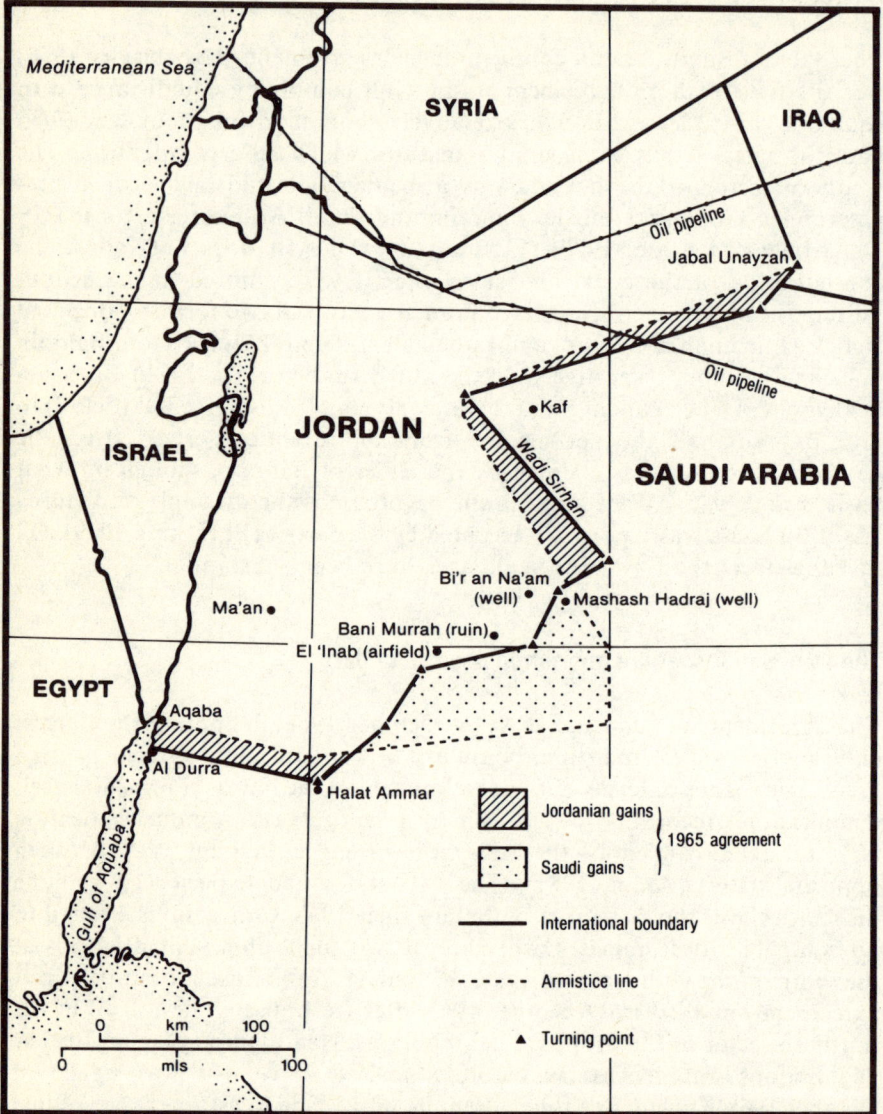

Figure 11.2 The Jordan-Saudi Arabia boundary (After U.S. Department of State, *International Study 60: Jordan-Saudi Arabia*, 1965).

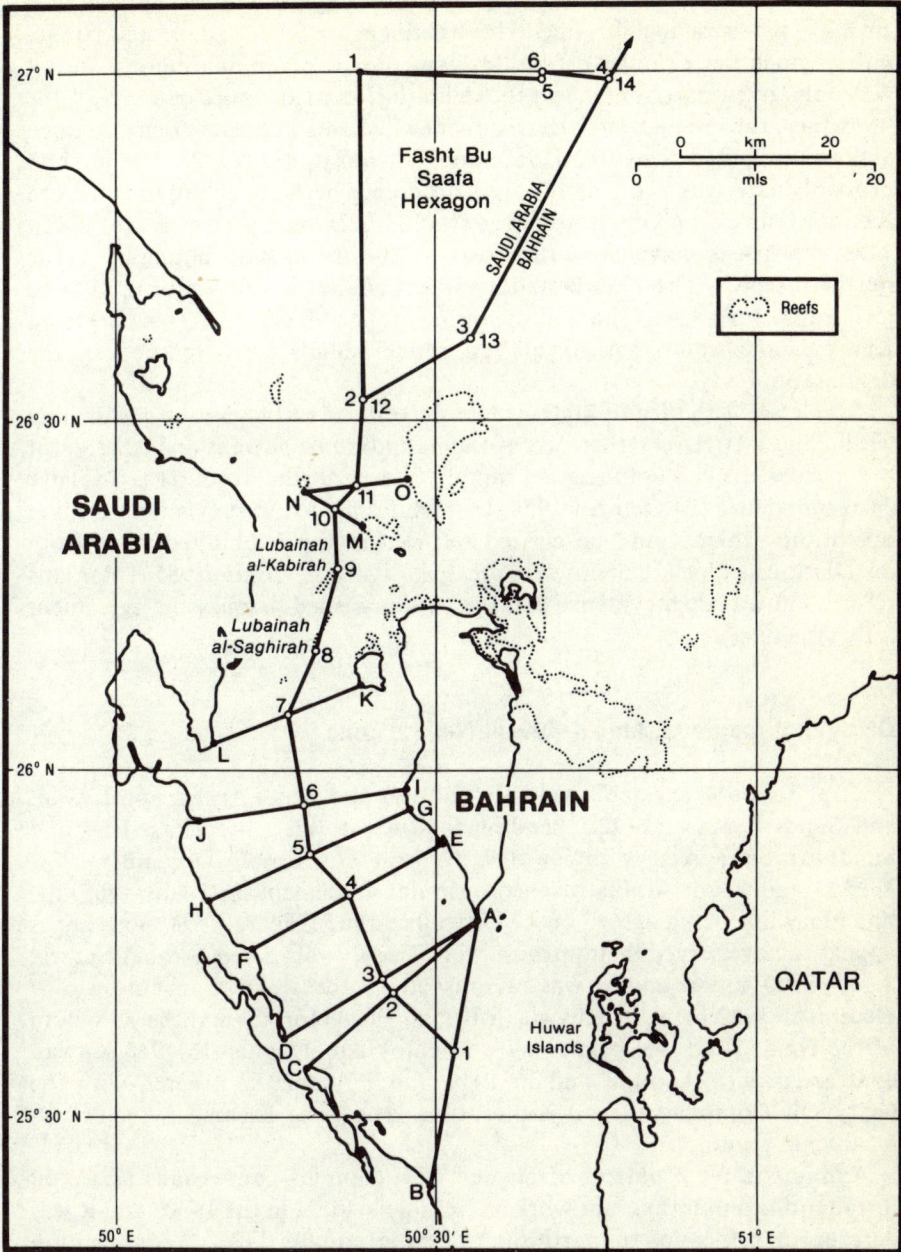

Figure 11.3 The Bahrain-Saudi Arabia boundary (After B. Conforti and G. Francalanci, *Atlante Dei Confini Sottomarini*, Milan, 1979).

Iranian and Al-Arabiyah as Saudi Arabian. The islands were allocated 12 mile territorial seas. Being only 13 miles apart they were thus separated by a curious east–west median line. The treatment of Kharg island (12 square miles) which lies 17 miles off the Iranian coast is of special interest since it was only given partial effect in the delimitation of the northern end of the boundary, thus predating such treatment of islands in later settlements such as France/United Kingdom (1982) and Libya/Tunisia (1982). The method probably used was to divide the areal difference between giving full effect to Kharg island and giving it no effect at all (US Department of State, 1981). This appears to have been the basis of the draft 1965 boundary in its northern sector. The 1965 boundary was modified in this sector in 1968 to give Iran a 'far greater' share of the resources (Amin, 1981: 103). Apart from Kharg, Farsi and Al-Arabiyah, all other islands were ignored in the delimitation.

Articles 4 of the Iran-Saudi Arabia agreement forbids oil or gas drilling within 500 metres on either side of the boundary to prevent one or other of the parties from capturing an unfair share of the resources. In later correspondence (24 October 1968) both states agreed to prevent 'all drilling operations which could be carried out within the Prohibited Area from installations which are themselves located outside it' (Amin, 1981: 151). This refinement was clear evidence that the parties wished to make the agreement fully effective.

Division of the Saudi Arabia-Kuwait Neutral Zone

Neutral Zones were established between Iraq and Saudi Arabia and Kuwait and Saudi Arabia in 1922, pending final territorial settlements. Iraq and Saudi Arabia agreed to divide their Neutral Zone on 26 December 1981. Kuwait and Saudi Arabia reached a similar agreement on 7 July 1965, the boundary line being agreed on 17 December 1967 (Figure 11.5). Both zones appear to have served their purpose well. The Kuwait–Saudi Arabia Neutral Zone (2500 square miles) was established by the Uqair Convention of 2 December 1922 between Kuwait (a British Protectorate) and the Kingdom of the Nejd. Here, the parties were to enjoy equal rights. In 1948 Kuwait contracted with Aminoil and in 1949 Saudi Arabia contracted with the Getty Oil Company for oil exploration. The zone became a major oil-producing region.

Administrative problems associated with complex concessions areas, oil installations and immigrant workers became evident in the 1950s and it was agreed, in principle, to partition the zone equally. The 1965 partition agreement gave the parties full rights of administration, legislation, and defence in their own parts of the Partitioned Zone (as it was to be called), at the same time preserving the rights of both parties to the natural resources of the whole Partitioned Zone (Article 3). Thus, existing oil concessions and

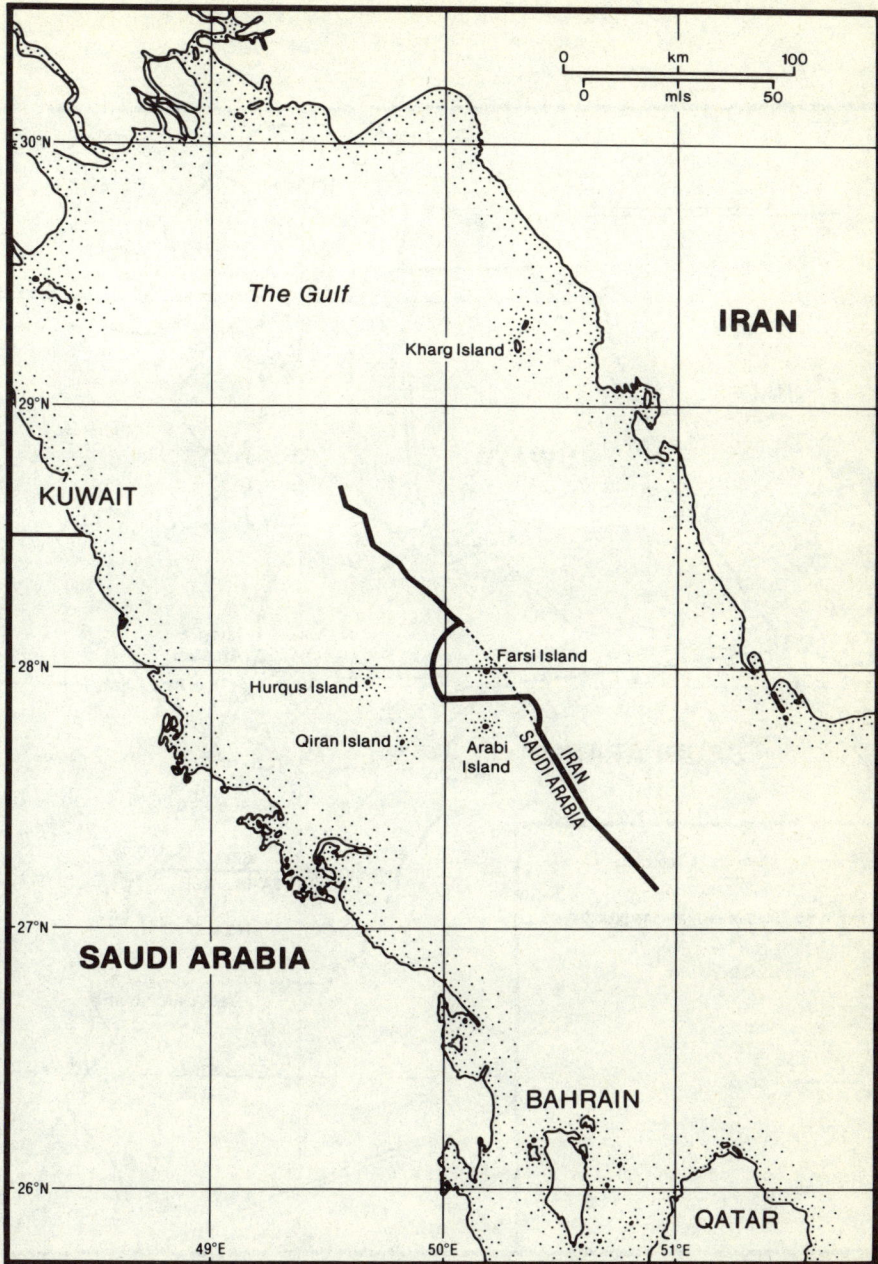

Figure 11.4 The Saudi Arabia-Iran boundary (After International Court of Justice, *Libyan Counter Memorial*, Vol. 2, The Hague, 1983).

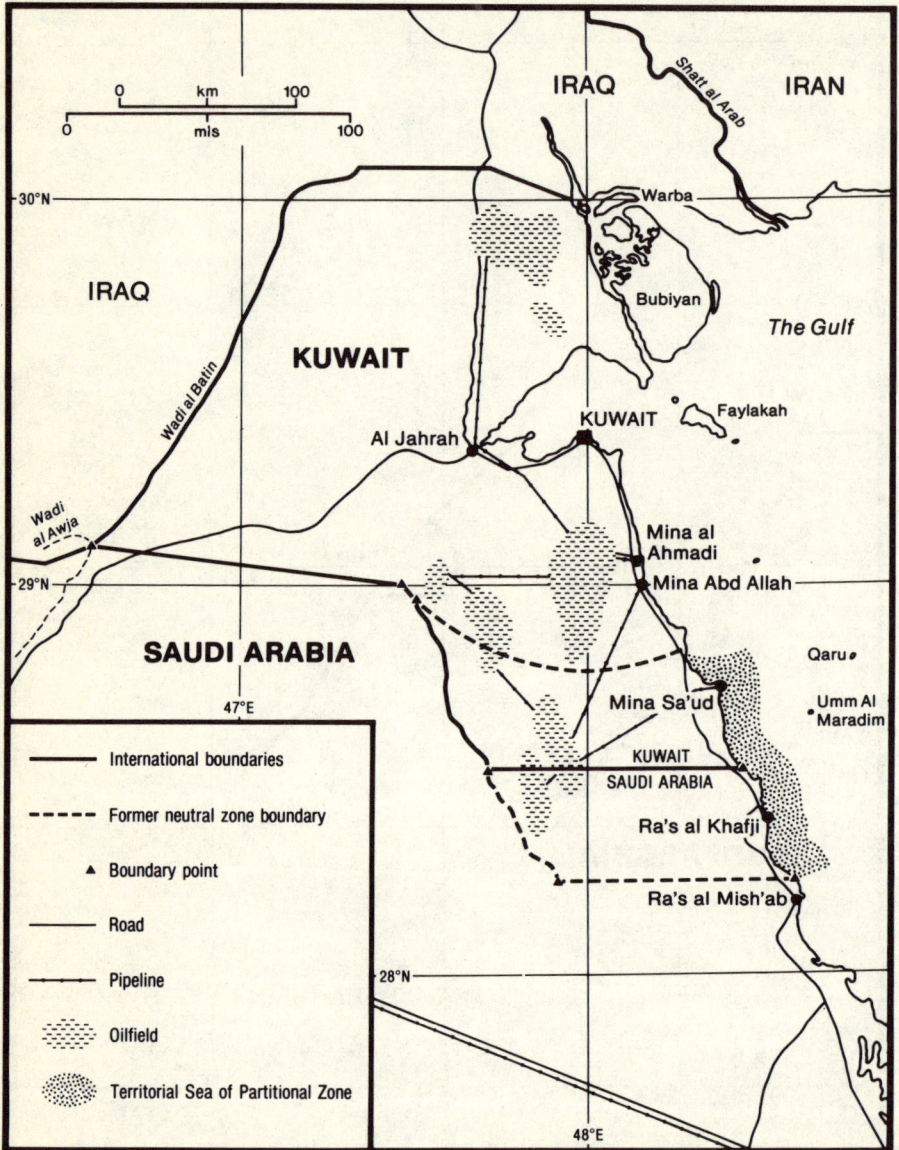

Figure 11.5 The Kuwait-Saudi Arabia boundary (After U.S. Department of State, *International Boundary Study 103, Kuwait-Saudi Arabia*, 1970).

production arrangements were unaffected. Each state also guarantees freedom for the citizens of the other to work in its own annexed territory (Article 15) (Day, 1982). The entire 101-mile Kuwait–Saudi Arabia boundary is demarcated in desert territory. The Neutral Zone partition line is a straight line 33.5 miles long (US Department of State, 1970).

The partitioning of the Neutral Zone finalized the shape of the State of Kuwait – which might have been three times larger but for Britain's concession to Ibn Saud at the Uqair conference in 1922. Britain's role had been deeply resented by the Sheikh of Kuwait who spoke of renouncing an 'unjust frontier' and recovering 'lost territories' (Al-Mayyal, 1986). The partition agreement did not finalize offshore arrangements, which remain a potential source of tension between Kuwait and Saudi Arabia. The rights of each state apply to the territorial waters adjacent to the Partitioned Zone to a distance of six nautical miles, although Kuwait and Saudi Arabia both claim territorial waters of 12 nautical miles elsewhere (Articles 7 and 8). Beyond six miles the area has not been partitioned, largely because of the disputed status of the small islands, Qaru (23 miles offshore) and Umm al-Maradim (16 miles offshore). Thus, beyond the six-mile limit Kuwait and Saudi Arabia continue to exercise their original rights in an area of seabed which has not been partitioned. Kuwait might be willing to share oil revenues accruing from the vicinity of Qaru and Umm al-Maradim in return for recognition of exclusive Kuwaiti sovereignty over the islands (Amin, 1981: 128).

Conclusion

Four of the five agreements discussed above represent peaceful solutions to potentially serious boundary conflict (Table 11.1). The Jordan-Saudi Arabia agreement is an exception. Although Ibn Saud had never formally relinquished his claim to territory around the Gulf of Aqaba, it had not been actively pursued and the risk of conflict was remote. The significance of the Treaty of Amman lies in the creative exchange of territory which occurred. This was relatively simple because the territories involved were sparsely populated or uninhabited, there were no known hydrocarbon deposits and the agreement was between neighbouring Arab monarchies enjoying cordial relationships.

Such preconditions are rare elsewhere in the Middle East and North Africa and territorial trading is unlikely to be widely adopted to rationalize colonial land boundaries. Similarly, the Neutral Zone partition agreement was between friendly Arab monarchies in a region where the practical arguments for partition were pressing. The most instructive aspect of the Neutral Zone agreement was the willingness of the parties to leave the tricky problem of disputed islands for later solution, while reaching settlement on non-controversial matters.

Table 11.1   Boundary agreements cited: a summary

Agreements	Territorial pre 1930	Territorial (contemporary)	Resources
Land Boundaries			
Jordan/Saudi Arabia (1965)	—	✓	✓
Neutral Zone Partition (1965)	—	✓	✓
Maritime Boundaries	Island Sovereignty	Resources	
Red Sea Common Zone (1974)	✓	—	
Bahrain/Saudi Arabia (1958)	—	—	
Saudi Arabia/Iran (1968)	—	—	

As to offshore agreements, the Red Sea Common Zone is interesting only in as far as it demonstrated a flexible approach to offshore delimitation. The precise geographical and political conditions are unlikely to be found outside the Red Sea, though the principles are similar to those in the 1982 UN Convention on the Law of the Sea for the exploitation of the seabed beyond national jurisdiction. The Common Zone could be adopted by other Red Sea states, although the resource incentives to do so are currently rather weak. It was reported in 1988 that North and South Yemen had agreed on a joint petroleum development zone in formerly disputed territory.

The Saudi Arabia–Bahrain and Saudi Arabia–Iran agreements demonstrate potentially useful approaches to maritime boundary delimitation. In both cases, disputed islands and resources, and the role of islands in boundary delimitations were major issues. The solutions depended on a combination of equidistance and equitable principles, and the acceptance of a number of compromises applied to a single boundary delimitation. As shown elsewhere (Blake, 1987), the use of such solutions as common zones or shared revenues is increasingly common in the delimitations of maritime boundaries throughout the world. In several important cases (for example, France/United Kingdom, 1982; Libya/Tunisia 1982) large islands have been given half or partial effect in boundary alignments. These methods were pioneered in the Gulf 20 or 30 years ago and used ingeniously in combination with other methods. It is this which makes the Gulf agreements so important. Moreover, the attempt to delimit the Iran–Saudi Arabian boundary so as to allocate known oil reserves fairly was also a bold and unusual technique. Possibly because of the provision of a 500-metre Prohibited Zone, it appears to have worked successfully.

On the basis of these examples it may be possible to identify five preconditions for the peaceful resolution of boundary disputes. First there must clearly be the political will to settle; in all the above agreements Saudi diplomacy played a major part. Second, there must be some incentives to settle such as an advantageous territorial exchange or the opportunity to exploit resources. Third, there needs to be some symmetry in the bargaining counters, such as two islands, or equal areas of land, or seabed owned by both parties. Fourth, the absence of a dense population greatly helps. Finally, political flexibility, in the form of the willingness of the parties to adopt untried ideas, reach partial agreements, combine methods, or make concessions, is needed. Unfortunately, in too many other Middle Eastern cases such preconditions do not yet exist.

References

Amin, S. H. (1981) *International and Legal Problems of the Gulf*. Wisbech: Menas Press.

Blake, G. H. (ed.) (1987) *Maritime Boundaries and Ocean Resources*. London: Croom Helm.

Blissenbach, E. and Nawab, Z. (1982) 'The Atlantis II deep deposits of the Red Sea', *Ocean Yearbook*, **3**, 77–104.

Brownlie, I. (1979) *African Boundaries*. London: C. Hurst and Co. pp. 111–20.

Day. A. J. (ed.) (1982) *Border and Territorial Disputes*. Longman (Keesing's Reference Publications), pp. 230–31.

El-Hakim, A. (1979) *The Middle Eastern States and the Law of the Sea*. Manchester: Manchester University Press.

Lay, S. H., Churchill, R. and Nordquist, M. (eds.) (1973). *New Directions in the Law of the Sea*, Vol. 1. New York: Dobbs Ferry, Oceana, p. 119.

Marston, G. (1984) 'Potential legal problems in the Red Sea'. In A. M. Farid (ed.) *The Red Sea: Prospects for Stability*. Beckenham: Croom Helm.

Al-Mayyal, A. (1986) *The Political Boundaries of the State of Kuwait: a Study in Political Geography*. Ph.D Thesis, School of African and Oriental Studies, University of London.

Mustafa, Z. (1984) Red Sea Resources. In A. M. Farid (ed.) *The Red Sea: Prospects for Stability*. Beckenham: Croom Helm.

Schofield, R. and Blake, G. H. (eds) (1988) *Arabian Boundaries: Primary Documents 1853–1957, Vol. 5*. Archive Editions.

US Department of State (1965) *International Boundary Study No. 60: Jordan–Saudi Arabia Boundary*. Washington, DC: US Department of State.

US Department of State (1970) *Limits in the Seas No 12: Continental Shelf Boundary: Bahrain-Saudi Arabia*. Washington, DC: US Department of State.

US Department of State (1970) *International Boundary Study No. 103: Kuwait–Saudi Arabia Boundary*. Washington, DC: US Department of State.

US Department of State (1981) *Limits in the Sea No. 94: Continental Shelf Boundaries: The Persian Gulf*. Washington, DC: US Department of State.

12 Apartheid as Foreign Policy: Dimensions of International Conflict in Southern Africa

Anthony Lemon

A destabilized regional sub-system

Geographers are seldom negotiators, but they can contribute to conflict resolution both by analysis of past successes and failures, and by greater understanding of the context and nature of present conflicts. Blake (this volume Chapter 11) does the former, chronicling the predominantly successful resolution of Arabian boundary disputes. The territorial integrity of the South African state is not seriously in question, notwithstanding occasional academic proposals for partition (see Maasdorp, 1980; Lemon, 1987). Only its continued control of the Namibian exclave of Walvis Bay, and its attempted internal 'decolonization' by granting 'independence' to African 'homelands', are controversial. However the resettlement by the white minority government of over 3 million Africans in those territories since 1960 dwarfs the resettlements of minorities described in this volume (Chapter 13) by Rumley, Carmon and Yiftachel in Malaysia and Israel.

Both Israel and South Africa are dominated by ethnic groups which are beleaguered minorities within their respective regions. In Israel both the boundaries of the state (and even its existence) and its constitution are at stake (see Kipnis, this volume Chapter 15). In South Africa the struggle is for power within existing boundaries but, as in Israel the conflict has long assumed a regional dimension; the controlling groups in both states have, to say the least, been aggressive in their own defence.

The global significance of southern Africa is not immediately obvious. The combined population of South Africa and Southern African Development Coordination Conference (SADCC) is only 106 million, of which South Africa accounts for one-third; its area is just under 7 million sq. km, roughly that of Australia (Figure 12.1). The region's total armed forces are about equal to those of Indonesia, or to a quarter of those of Vietnam (Martin, 1988: 1). Yet no other Third World region apart from the Middle East attracts greater international attention.

The reasons for this are not primarily related to the region's role as a 'treasure house' of gold and strategic minerals, and still less to the strategic

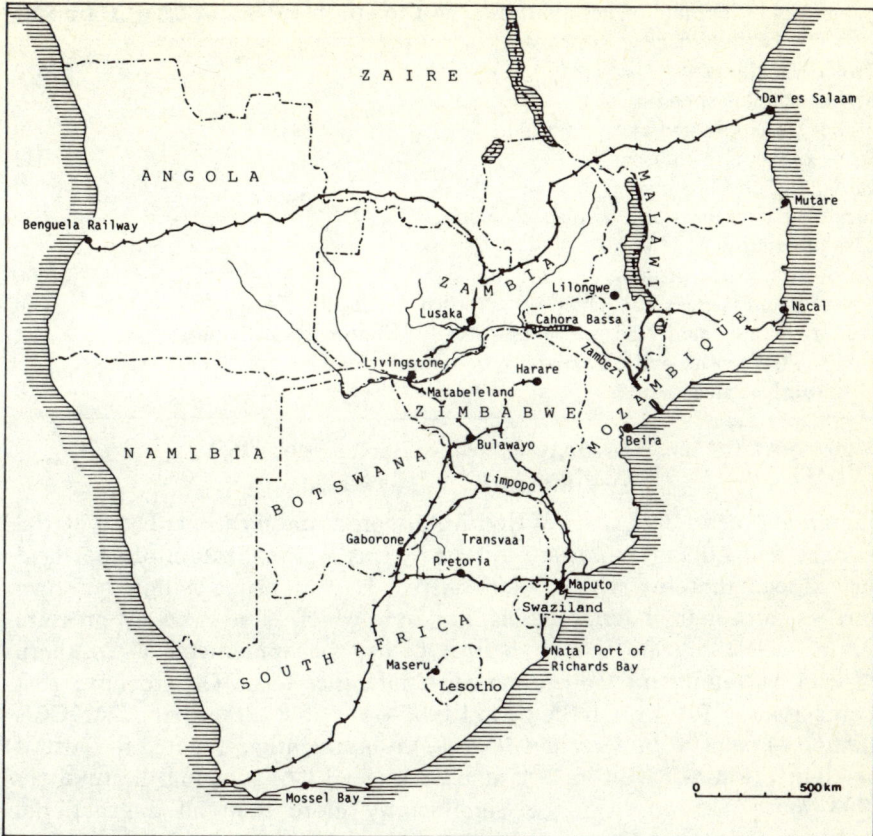

Figure 12.1 South Africa – geopolitical setting

value of the Cape sea lanes. Far more important is the persistence of apartheid in South Africa, and the regional ramifications of this. Pretoria's insistence on ethnicity as the major organizing principle of society, as a means of continuing white political domination, is offensive to liberals and radicals alike. South African and southern African economic and geo-graphical realities vividly mirror global north–south relationships of wealth and poverty, while their association with a broadly capitalist system in which the West has a longstanding involvement also gives the situation its East–West dimension.

What is just offensive to the international community at large is an anathema to South Africa's neighbours, who share a common interest with the black majority in South Africa and the liberation movements outside in ending white minority rule. Pretoria's determination to sustain it has led to a variety of responses – economic, political and military – which are the subject of this chapter. International concern has focused above all on the

Table 12.1   Estimate of destabilization cost to member states of SADCC (in $m)

Direct war damage	1610
Higher defence spending	3060
Higher transport and energy costs	970
Lost exports and tourism	230
Smuggling and looting	190
Supporting refugees and displaced persons	660
Loss of existing production	800
Loss of economic growth	2000
Boycotts, embargoes and sanctions imposed by South Africa	260
Trading arrangements (losses which, SADCC claims, result from Botswana, Lesotho and Swaziland remaining in the Customs Union with South Africa)	340

Source: SADCC report to OAU, quoted by Hanlon (1986: 265).

human and economic costs of South African destabilization. The material damage and human suffering inflicted by this regionalization of apartheid far exceeds that for which South Africa is responsible within her own borders, although it tends to be the latter which is stressed by pressure groups. Destabilization was estimated by the nine-member Southern African Development Co-ordination Conference (SADCC) to have cost member states $10.1 bn in the period 1980–84 (Table 12.1). Using SADCC's method of calculation Green *et al.* (1987) subsequently estimated the cost of destabilization at $7 bn in 1985 and $8 bn in 1986, bringing the total for 1980–86 to $25 bn, which is significantly more than all international development assistance in the same period.

Hanlon (1987: 339–40) has estimated that 735,000 people died as a direct or indirect result of destabilization in the same period. This includes an estimated 150,000 victims of war and war-induced famine in Mozambique, and a further 50,000 in Angola. The remaining deaths were, Hanlon claims, those of over half a million children in the two countries from the destruction of health facilities and especially the curtailment of immunization programmes.

Such estimates must be controversial, and the degree to which South Africa alone is responsible is impossible to quantify: low-level conflicts would undoubtedly continue in Mozambique and Angola, for example, even without Pretoria's involvement. But the international community correctly perceives South Africa as the chief architect of regional instability, and this perception is unlikely to change fundamentally while white minority domination continues.

The objects of South Africa's regional policies are less clearly understood; they are commonly oversimplified as maximizing damage to the national economies and social structures of neighbouring states with a view to weakening their political systems and in some cases overthrowing their governments, thus increasing Pretoria's economic, political and military dominance over the region. However such policies are potentially in conflict with other prime objectives of the apartheid state including the regional interests of South African capital and the desire both to avoid an escalation of international economic sanctions and to minimize Pretoria's international isolation. These interests would arguably be served by more positive regional relationships with economically stronger and politically stable neighbours, and the promotion of such relationships has indeed been an important strand of South Africa's regional policy.

The apparent conflict between these two sets of objectives has led to the popular perception of a fluctuating balance between the 'hawks', desiring to keep South Africa's neighbours cowed, and the supposed 'doves' in the Cabinet, who want South Africa to use her strong economic base to buy stability. Such policy conflicts are seen to be revealed when diplomatic negotiation on one front is threatened or negated by military action on another, as when the 1982 commando raid on Maseru, Lesotho, coincided with negotiations between South African and Angolan diplomats in the Azores (Lodge, 1983: 52), or when the 1986 raids on alleged ANC targets in Zambia, Zimbabwe and Botswana effectively scuppered the mediation efforts of the Commonwealth's Eminent Persons Group.

Such a view distorts the real situation. Hawks and doves do exist, and there have been serious disagreements on specific issues, though these should not invariably be interpreted as clearcut power struggles between the South African Defence Force (SADF) and the Department of Foreign Affairs. Such disagreements have usually concerned tactics in specific situations and are underlain by a broad congruence of strategic aims and a logical connection between superficially contradictory policies (Davies and O'Meara, 1987).

Destabilization has never been a first option for South Africa: it is a means to an end. Pretoria's ideal option would be full diplomatic political security and economic relations with friendly governments in every regional state (Martin, 1988: 33). Its broader strategies aim to secure its neighbours' compliance with this scenario, or at least with some elements of it. It is usually when these strategies fail and South Africa's regional policies are in crisis that destabilization becomes a more prominent element of policy. Whatever its success in browbeating neighbouring states into submission, destabilization has major costs for Pretoria; the greater its extent and duration, the more serious the damage to the pursuit of regional economic and international political goals. Quite simply, waging war is not conducive to the peaceful coexistence (on its own terms) for which South Africa ultimately strives.

Vale (1987) conceptualizes regional co-operation in terms of a spectrum of approaches, from functionalism to assertive incorporation to coercive integration. The functionalist approach sees states following self-interest in matters of co-operation, making arrangements for their mutual benefit which will first grow into habits of behaviour and later into institutional structures (Banks, 1969: 346). The Southern African Customs Union (SACU) agreement of 1910, between the newly formed Union of South Africa and the British Protectorates of Bechuanaland, Basutoland and Swaziland, exemplifies the functionalist approach.

However the renegotiation of SACU in 1969, following the independence of Botswana, Lesotho and Swaziland, was much more reflective of an assertive incorporation defined by Vale (1987: 182) as 'a regional peace which would tolerate apartheid in return for the advantages of co-existence'. The weaker states in the region had little or no choice but to accept some degree of assertive incorporation, although for two – Swaziland and Malawi – a degree of functionalism also characterizes their relationships with Pretoria. The abortive 'constellation of southern African states' which is discussed below represents an attempt by Pretoria to push assertive incorporation further than the region's geographical realities forced its neighbours to go. Even the use of what Pretoria euphemistically calls 'transport diplomacy' has failed to secure this regional policy prize.

When assertive incorporation fails, coercive integration follows. This is essentially military, and involves direct attacks on neighbouring states, support for dissident movements, and semi-conventional warfare as in Angola. But Pretoria has also used its economic power coercively, most obviously in the virtual blockade applied to Lesotho in January 1986, which led to the fall of the government.

The techniques of coercion: (1) economic weapons

South Africa is a semi-peripheral state or a secondary developing core in the world economy. It has a subordinate relationship with the major industrial countries but a superordinate relationship with the peripheral Third World states around it, which it can exploit. Within the constraints of the world economic system, only limited possibilities exist for peripheral Third World states to advance (Wallerstein, 1974). They are effectively forced to choose between dependent development and a strategy of revolutionary trans- formation, which involves greater self-reliance and an attempt to build completely new structures.

The economic and political interests of apartheid demand that the states of southern Africa embrace dependent development, especially in regional terms. Those which have attempted revolutionary transformation, notably Angola and Mozambique, and those such as Zimbabwe which incline in the same direction but are more constrained by the inherited structures of settler or colonial economies, are made forcibly aware of both the structural constraints on transformation and the economic, political and military price of attempting it.

Material dependence on South Africa revolves primarily around migrant labour, transport and communications, energy and industrialization. South

Africa's use of these has been analysed in detail elsewhere (especially Hanlon, 1986; Martin, 1988), while Geldenhuys (1981) has given detailed consideration to 'the ways in which the Republic could use its economic relationships in southern Africa for non-economic purposes'.

For Mozambique, Botswana, Swaziland and above all Lesotho, South African jobs are a major source of national revenue, especially since the dramatic increases in mineworkers' wages in the mid-1970s. South Africa has repeatedly used labour dependence for political ends. It has threatened to expel migrants and end recruiting if subjected to economic sanctions, although a total ban would be resisted by the mining companies. When Lesotho refused to follow Mozambique in signing a non-aggression pact in 1984, South Africa threatened to transfer some of Lesotho's migrant quota to Mozambique, but did not do so. In October 1986 Pretoria implemented a threat to expel Mozambiquan workers from South Africa in retaliation against Mozambique's alleged harbouring of ANC guerillas. However the Chamber of Mines subsequently negotiated an amelioration of the original decision which will result in skilled workers keeping their jobs, but no novices will be recruited (Grest, 1987: 361).

In many respects, South Africa's industrial dominance is continental as well as regional. However, many of its goods remain relatively expensive because of the small market for which they are produced and inefficiency induced by protectionism. To compete with goods from other countries South Africa resorts to a range of export incentives including generous credits, as well as unusually favourable long credit terms for importers, with which SADCC producers cannot compete (Hanlon, 1986: 71–2). South African goods also enjoy the advantage of proximity compared with Western goods. Moreover the dramatic decline in the exchange value of the rand since 1985 has made South African goods considerably more competitive in the region. This has adversely affected the small and relatively weak manufacturing sectors of its neighbours, including Zambia (until its dramatic devaluation in late 1985), Botswana, Malawi and especially Zimbabwe, which has by far the most developed manufacturing sector in SADCC.

The mining industries of Zambia and Zimbabwe are geared to the import of South African equipment. Attempts to switch to other suppliers would increase costs, and require the mining companies to secure foreign exchange in advance for a major extension of delivery lead times (Martin, 1988: 23).

Several states display varying degrees of energy dependence on South Africa. Lesotho is completely dependent on South Africa's Electricity Supply Commission (ESCOM) grid, while Mozambique imports 44 per cent of its power from South Africa, all for the Maputo area. Swaziland has a large power potential, and has decreased its dependence on ESCOM to 30 per cent, and Botswana now uses ESCOM only as a back-up supply since completing its own coal-fired central power station and electricity grid in 1986.

While Angola produces oil, Botswana, Lesotho and Swaziland buy all their oil products from South Africa. Malawi buys most of its fuel from South Africa. Some of Zimbabwe's needs are imported through Maputo and railed via South Africa. Pretoria has gone to considerable lengths to maintain its control over the region's fuel supply, refusing to supply Lesotho with sufficient fuel for a planned four-month strategic reserve in 1980 and refusing to rail Algerian fuel from Maputo to Lesotho in 1982 for the same purpose (Hanlon, 1986: 74–5). This strong stand is clearly related to the sensitivity of oil supplies for South Africa itself in the face of sanctions; although oil-from-coal plants and the liquefaction of natural gas from Mossel Bay will make South Africa increasingly self-sufficient (Lemon, 1987: 148–9), it realizes the hostage value of its potential stranglehold over energy supplies to weaker, landlocked neighbours.

Energy also provides a limited element of reverse dependency. When operative, the Cahora Bassa hydro-electric plant in Mozambique can supply 8 per cent of South Africa's electricity needs; the power lines have been constantly attacked by the Mozambique National Resistance (MNR), but in January 1988 South Africa and Mozambique agreed to contribute to the rehabilitation of the transmission lines and the provision of security forces to guard them.

In only one case has South Africa embarked on a project which makes her dependent on a black-ruled state. This is the Lesotho Highlands Water Project which had been under discussion for many years but was finally agreed only in February 1986, just one month after the Lesotho coup, catalysed by an effective South African blockade, which brought to power a government much more sympathetic to South Africa. The first water is scheduled to flow in 1995, in time to prevent the southern Transvaal experiencing serious water problems. The project comes as close as anything in the region to the 'functionalist' category of relations between South Africa and a SADCC member state. It will cement still further the economic integration of Lesotho with South Africa, while giving Lesotho slightly more leverage in the relationship, and making it self-sufficient in electricity.

Transport is the most critical aspect of dependence, and also both the direct economic weapon Pretoria has used most frequently and the principal focus of its destabilizing activity in Mozambique. The six landlocked countries in the region constitute 'a periphery within a periphery' (Reitsma, 1980). Lesotho is entirely surrounded by South Africa, and thus at Pretoria's mercy, as the January 1986 blockade showed only too well. Swaziland has rail access to Maputo; this line, although periodically attacked by the MNR since 1984, most recently in January 1988, has never been definitively closed. Despite the opening of a line to the Natal port of Richards Bay in 1978, the Maputo line continues to take two-thirds of Swazi exports. Botswana has traditionally depended on South African rail routes, and lacks a practical alternative.

Zambia, Zimbabwe and Malawi were dependent on rail outlets through

Mozambique, not South Africa, in colonial times: the lines to Beira and Maputo, and to Nacala which was extended to Malawi's new capital at Lilongwe in 1970. The use of South African routes was enforced by the Mozambique border closure with white-ruled Rhodesia in 1976. Continuation of this pattern has been ensured by repeated MNR attacks on all three rail routes, the rehabilitation and securing of which are priority projects for SADCC, and have attracted substantial Western aid.

Paradoxically it is Malawi, the only state in the region to have diplomatic relations with South Africa, which has suffered most; the Nacala line has been closed since 1984, forcing almost all Malawi's traffic to take the long and costly southern route by road and rail. In addition to current attempts to rehabilitate and protect the Nacala line, a project to facilitate traffic in Malawi's northern corridor, by road to Lake Malawi ports and by steamer to the Tazara railway at Mbeya, is under way. While the viability of such double trans-shipment looks doubtful, the potential security of the northern route appears high.

Rehabilitation of the Beira line is SADCC's leading project (Martin, 1988: 84). The line was closed by MNR attacks between 1975 and 1980, and since then has run at very low capacity and usage rates. Forecasts depend on assumptions about security, spending rates on physical infrastructure and management quality, but the line should be carrying one-third of Zimbabwe's non-oil tonnage by the early 1990s. If Mozambique would cede control of the whole line to Zimbabwe, progress would probably be more rapid, but this appears unlikely (ibid., p. 85). Meanwhile the port of Beira remains vulnerable to SADF attack; it destroyed channel marker buoys in 1981, and the tank farm serving the oil pipeline to Mutare, Zimbabwe in 1982 causing a brief but acute fuel crisis.

Since 1987 the Limpopo line to Maputo has increasingly been the focus of SADCC attention. This was traditionally Zimbabwe's main outlet, and was also more attractive than Beira for Zambian and Zairean traffic via Bulawayo. Maputo port is bigger and better equipped than Beira, and the line is flatter, straighter and faster. For Mozambique itself the line is a top priority as the main artery of Maputo Province. Its drawback is security, as most of the line runs within 100 km of the South African border. Maputo port is also badly run down, and essential dredging has been neglected as traffic has dwindled from 14 million tons in 1973, half of which served South Africa and Swaziland, to only 2.5 million tons in 1987. South African Transport Services (SATS) have deliberately diverted much traffic away from Maputo, against the wishes of their own exporters. A major rehabilitation programme is now under way, but has lower priority and is less carefully co-ordinated than other SADCC projects, since South Africa is still the largest single customer (Martin, 1988: 89).

The Tanzam railway (Tazara) was completed by the Chinese in 1975, the year in which the still-closed Benguela line across the Shaba Province of Zaire and Angola was closed by UNITA. Tazara's problems have been its single-track line, with passing places, poor management and maintenance, and pilfering. It is, however, the most secure route in the region, and a ten-year rehabilitation programme is under way, coinciding with the upgrading of the port at Dar es Salaam. Zambia decided in December 1986 to discontinue all mineral exports through South Africa as soon as possible, and has already reached a substantial degree of transport independence, largely based on Tazara.

The techniques of coercion: (2) military means

The only large-scale army offensives have been in Angola and Namibia. Elsewhere direct attacks have been made by 'reconnaissance' (commando) troops, based in the Kruger National Park on the Mozambique border, the northern Transvaal and the Caprivi Strip. Raids into neighbouring states in the 1980s have used various means of surface, sea, and air transport. The commando battalions are also known to have contingency plans for attacks on all obvious economic targets such as bridges, oil facilities and power stations (Martin, 1988: 35).

In addition to sometimes serious economic damage, such commando raids are of major symbolic importance, acting as a constant reminder of Pretoria's willingness to use its military power and adding to the insecurity of SADCC states. Most material damage has, however, been inflicted by proxy through the support of dissident movements. The main victims have been Angola (UNITA) and Mozambique (NMR), but South Africa has also supported Super-Zapu in Zimbabwe's Matabeleland province and, until the January 1986 coup, the Lesotho Liberation Army.

Zambia has experienced relatively little military destabilization, except in the remote corner west of the Zambezi where frequent South African incursions, mine explosions and harassment of local people are related to the Angolan war. Otherwise the weakness of its economy and what Pretoria regards as the relative 'moderation' of Dr Kaunda are such that it is not viewed as a threat to South African hegemony.

Nowhere has South African-backed destabilization been greater than in Mozambique. Direct and indirect attacks on ANC premises have been more frequent than elsewhere, partly because Mozambique was the main infiltration route. Such attacks damaged prestige and morale, but the real damage has been done by the MNR, an organization adopted by South Africa in 1980, having been founded in the late 1970s with assistance from the white settler regime in Rhodesia. By 1987 it was thought to number around 20,000, about half of whom were trained soldiers. It has created no alternative government structures in the areas it controls, and only in a few areas such as the Zambezi valley does it appear to command popular support; elsewhere it has committed atrocities and carried out widespread intimidation and banditry, as well as destroying lines of communication and other infrastructure. It has no well known political leaders and no political programme beyond a generalized anti-Marxism, and is not seen as a credible alternative government, although in early 1986, after the MNR had recaptured the Gorongosa base from Frelimo, the latter did consider a power sharing compromise. No talks took place, however, in the absence of a credible assurance from the SADF that they would not see this as a stepping stone to a full assumption of power by the MNR (Martin, 1988: 50). In 1987, President Chissano publicly ruled out negotiation with the MNR as 'a product of colonialism and South Africa', but in mid-1989 talks were planned under the co-sponsorship of Presidents Moi (Kenya) and Mugabe (Zimbabwe).

A government assessment in 1988 revealed that the MNR had destroyed 1,800 health centres, 720 schools, 900 rural shops, 1,300 lorries and 44 agricultural centres, including major tea and sugar factories (KCA, 1988: 35685). An estimated 5 million people dependent on subsistence farming had been forced from their land. In late 1987 there was a marked increase in the number of MNR attacks on civilian transport on the main road and railway out of Maputo, with great loss of civilian life, and it was feared that the government was losing control of the main transport arteries.

Angola has no significant transport or other economic links with South Africa, which has used military means to fight SWAPO in both northern Namibia and southern Angola, and to support the UNITA movement of Dr Jonas Savimbi in its struggle for a share in governing Angola. Unlike the MNR, UNITA is a genuine liberation movement which fought the Portuguese and has a popular support base, particularly among the Ovimbundu people of southern and parts of central Angola.

South African support for UNITA continued, with the result that the MPLA government lacks *de facto* control over much of its territory. UNITA's role in keeping the Benguela railway closed has also been important to Pretoria. UNITA has established its own government structures in the territory which it holds most securely in the south-west, where its capital of Jamba is located. In mid-1987 President dos Santos claimed that since independence in 1975 the war had cost Angola some $12bn in terms of economic sabotage; 600,000 Angolans had been displaced, 60,000 killed, and 'a great number' had sustained permanent disabilities.

Unprecedented four-party talks between government delegations from Angola, South Africa, Cuba and the USA began in May 1988. The final agreement involves a Cuban withdrawal from Angola over a two-year period and the implementation of UN Security Council resolution 435 to bring about Namibian independence. The exclusion of UNITA from the negotiations leaves the future of the Angolan civil war uncertain, although a ceasefire was agreed in mid-1989. South Africa's insistence on retaining Walvis Bay, Namibia's only port and a South African naval base, poses problems for an independent Namibian government and will inevitably cause conflict.

South African regional policy and the search for a 'Pax Pretoriana'

How does Pretoria's use of economic weapons and military destabilization relate to the pursuit of a 'pax Pretoriana' which serves South Africa's regional economic and international political interests? An answer to this question must consider chronological fluctuations in the policy mix pursued, as well as policy differences towards particular states; it also requires assessment of the underlying unity of purpose or otherwise which has existed in government and military over the period since 1974.

As early as 1963 the then Prime Minister, Dr Hendrik Verwoerd, proposed a common market or commonwealth as a first step to the establishment of a regional free trade zone, and ultimately, perhaps, a more political grouping. The 1960s saw increasingly close South African links with white-ruled Rhodesia and the Portuguese colonies, together with increasing intervention in Bechuanaland, Basutoland and Swaziland as they approached independence.

B. J. Vorster, Dr Verwoerd's successor, attempted to initiate dialogue with African governments in the late 1960s, and succeeded most in Malawi, where the establishment of diplomatic relations in 1969 was followed by an exchange of official visits in 1970 and 1971. The OAU condemned the dialogue initiative as a 'manoeuvre' in 1971, but six states including Malawi and Lesotho voted against and five more including Swaziland abstained.

The Portuguese revolution of 1974 heralded the imminent collapse of South Africa's 'cordon sanitaire' and clearly demanded a reformulation of regional strategy. Angola and Mozambique became independent under Marxist governments in 1975, SWAPO was enabled to operate along the entire northern border of Namibia, the guerilla threat to Rhodesia on its eastern border with Mozambique was magnified, and the closure of that border to transit traffic in 1976 forced the Smith regime to direct all its trade via Botswana and South Africa.

Pretoria's rethinking took two forms, which illustrate the complementarity between the two strands of its regional policy which was to be a major characteristic henceforth. The defence budget was immediately increased substantially, growing by more than three times between 1973–4 and 1977–8. But diplomatic and economic initiatives continued; B J Vorster sought 'detente' in the region, proposing to bring together southern African states into a 'constellation of completely independent states'. This constellation concept was later revived by P. W. Botha. Mr Vorster saw the 'constellation' as involving mutually beneficial economic and political interdependence and providing a buffer of friendly states between South Africa and a largely hostile continent.

Until now Pretoria had managed to keep foreign and domestic policies reasonably distinct; detente reflected the state's confidence in its ability to maintain order at home and strengthen economic and political ties abroad (Spence, 1987: 156). This changed sharply in the mid-1970s. The South

African invasion of Angola in 1975 dramatically contradicted Pretoria's traditionally strong opposition to interference in the domestic affairs of sovereign states. It underestimated, not for the last time, the effects of its own pariah status on the public stance of other governments. Lacking support South Africa withdrew, in what was perceived in some quarters as a military defeat at the hands of Third World (Cuban) troops. The hesitant steps towards domestic reform which followed represented a response to both external and internal pressures, many of the latter reflecting the growing contradictions between Verwoerdian apartheid and the consequences of rapid economic growth since the early 1960s.

The combination of the Angolan debacle and South Africa's brutal repression in Soweto rendered further dialogue politically impossible for Zambia and its other proponents. South Africa's response to this crisis for 'pax Pretoriana' not surprisingly included a major escalation of military activity, but, as in 1974, it also included yet another attempt to secure regional economic co-operation. Thus while the Defence White Paper of 1977 launched the defence chiefs' call for 'total strategy', the following year saw the re-launch of the constellation concept (CONSAS) by P. W. Botha, in which support from the private sector for a new Southern African Development Bank to finance development projects in the member states of CONSAS was called for.

'Total strategy' involved the mobilization of economic, political and psychosocial as well as military resources both regionally and internally. Instead of attempting merely to influence individual decision-makers, henceforth regional policy would seek to change the objective environment where decisions were made (Davies and O'Meara, 1987: 246). The White Paper specifically called for action relating to transport, distribution and telecommunications to promote 'political and economic collaboration among the states of southern Africa': such assertive incorporation was clearly seen as a necessary element in bringing CONSAS to fruition.

The aftermath of the Angolan invasion and the Soweto unrest hardly seems a propitious time to have re-launched CONSAS. Pretoria's hopes of success hinged above all on its expectation of a victory for 'moderates', led by Bishop Abel Muzorewa, in the Zimbabwe independence election in 1980. Swaziland and Malawi's accession to CONSAS was confidently expected; if Muzorewa added Zimbabwe, Lesotho and Botswana would be virtually forced to join. Zaire could then be persuaded, leaving Zambia under strong pressure, whilst Namibia could hopefully be brought to an internal independence settlement under a compliant government which would also join. Thus the 'pax Preoriana' seemed truly within South Africa's grasp.

Robert Mugabe's victory in the Zimbabwe elections was a shattering blow to Pretoria's hopes. Even apart from this, Pretoria had again underestimated the antipathy of its neighbours towards co-operation with apartheid. The black-ruled states of the region reacted by forming SADCC in 1980 with the object of seeking a new development strategy for the region to promote regional economic co-operation amongst themselves in the interests of accelerated social and economic development, greater self-reliance and reduced economic dependence. Even Swaziland and Malawi joined what amounted to a 'counter-constellation', which South Africa presented as a hostile alliance to justify her subsequent regional policy.

This was not wholly true, as early SADCC documents actually acknowledge an interest in preventing the imposition of mandatory economic sanctions against South Africa, recognizing that their own links with Pretoria would have to be maintained for a long time (Jaffee, 1983: 26). There was even some support in South African business circles for SADCC, believing that a stronger regional base would expand South Africa's markets and improve infrastructure, relieving bottlenecks in northward movement.

Such was not the majority view in government or the SADF. Assertive incorporation had temporarily failed, so Pretoria turned to coercive integration. There followed a period of generalized and seemingly indiscriminate military destabilization from mid-1980 to the end of 1981 in surrounding states. Economic pressures were also accentuated: more than 20 locomotives were withdrawn from Zimbabwe just when they were needed to cope with a record harvest and increased trade following the end of sanctions, and Pretoria threatened to end the preferential trade agreement between the two countries (but ultimately renewed it in July 1986).

A period of more intensive but also more selective destabilization began early in 1982, designed to frustrate SADCC efforts to reduce economic dependence and to limit the numbers and activities of the ANC, which was presented as a wholly external threat justifying South African pressure in states which harboured it. South Africa's two main ideological enemies suffered the most. In Mozambique there was a further escalation of MNR attacks, as well as some direct SADF attacks and a partial boycott of Maputo port. Southern Angola was virtually occupied by South African forces by the end of 1982, as the SADF intensified the struggle against SWAPO and sought to raise the costs to Angola of supporting it.

Lesotho was another victim at this time, as Pretoria increased its pressure on Chief Jonathan to neutralize the ANC presence. Armed activity by the Lesotho Liberation Army escalated considerably, and the major SADF raid on Maseru took place in December 1982. Border restrictions were imposed on Lesotho in mid-1983, and threats made to repatriate migrant workers.

The diversified nature and underlying strength of the Zimbabwe economy offered some scope for reducing economic ties with South Africa. Measures to prevent this included sabotage of the Beira-Mutare oil pipeline and the partial blocking of oil imports via South Africa. On the military front, support was given to Super-Zapu dissidents, backed by the broadcasting of 'Radio Truth' to Matabeleland.

Destabilization was only employed where other means failed. Swaziland's conservative rulers shared essentially the same ideological world view as

Pretoria (Daniel, 1984: 231), and carrots were sufficient to secure South Africa's ends. Agreement on the building of a railway to Komatipoort, providing a new Swazi transit line for South African and SADCC traffic to Richards Bay and Durban, was reached in 1982, following a R50m. extra payment to Swaziland from SACU; the line was opened in 1986. Meanwhile negotiations proceeded concerning a South African offer to cede the Kangwane 'homeland' and the Ingwavuma district of northern KwaZulu to Swaziland, in return for the latter's acceptance of some 750,000 ethnic Swazis in South Africa as her own citizens. The appeal of such territorial growth and access to the sea, notwithstanding the problems of more than doubling Swaziland's nominal population, strongly appealed to King Sobhuza, and contributed to his government signing a non-aggression treaty which was only revealed two years later when Mozambique's Nkomati Accord made it more acceptable.

South Africa's intensive destabilization activities elsewhere in the region led by the end of 1983 to American pressure on South Africa and its neighbours for rapprochement, while the Soviet Union warned Pretoria that it would not tolerate the fall of the MPLA in Angola. Meanwhile the costs of Operation Askari shocked both government and white public opinion in South Africa. Mozambique's situation was increasingly desperate, following both severe drought and recent flooding. These and other factors combined to offer yet another chance of securing the 'pax Pretoriana', and for a few months Western governments were encouraged by the progress made in the form of the Lusaka Agreement of February 1984 with Angola and especially the Nkomati Accord of March 1984 with Mozambique.

Under the Lusaka Agreement South Africa agreed to withdraw its troops from Angola in return for the latter's undertaking that neither SWAPO nor Cuban forces would move into the areas vacated by the SADF. The short-term effect on SWAPO was considerable, but South Africa did not complete its withdrawal until April 1985 – a year later than agreed – only to have a commando unit captured the following month engaged in sabotaging the Cabinda oil installations.

The Nkomati Accord was much more far-reaching. South Africa agreed not to support the MNR in exchange for Mozambique agreeing to prevent the ANC's armed wing, Umkhonto we Sizwe, operating from its soil; other provisions covered economic co-operation. Mozambique saw the agreement as isolating the MNR from its principal external backers and laying the basis for the conduct of regional relations in terms of international law. Pretoria likewise hoped for a significant downturn in the armed struggle when the ANC was deprived of its Mozambique bases; that this did not occur (during 1985 ANC activity increased sharply) reflected the strength of the covert ANC presence within South Africa.

For Pretoria the Accord had wider significance. Its insistence on the economic clauses reflected the hope of playing a leading role in the process of political, economic and social restructuring in Mozambique, moving the country away from socialism. Some of Pretoria's strategists clearly saw this as leading to the inclusion of the MNR in a power-sharing government. South Africa also hoped that Nkomati would set a precedent for other security agreements with its neighbours, and that it would allow Pretoria to claim international recognition as the *de facto* regional power in southern Africa (Davies, 1987).

These hopes enjoyed some initial success. P. W. Botha's June 1984 visit to Western Europe would have been inconceivable a year earlier. South Africa was to play a role in the implementation of a ceasefire in Mozambique; the ceasefire, however, failed to materialize. Botswana, Lesotho and Zimbabwe all reiterated that they would not permit liberation movements to operate from their territory but resisted South African pressure to sign ceasefire agreements.

The fusion of domestic and foreign policy was again apparent as the unrest which began in the Vaal Triangle in September 1984 spread rapidly, leading to the introduction of a state of emergency in the East Rand and the eastern Cape in July 1985. Two events in August contributed further to the destruction of the international goodwill briefly enjoyed after Nkomati. One was the capture of documents proving South African violations of Nkomati which had become increasingly obvious. The other was the disappointment of widespread international expectations of P. W. Botha's much heralded 'rubicon' speech.

The next phase is by now familiar. With the 'pax Pretoriana' once more in ruins, South Africa again resorted to coercive integration. Military aid to UNITA was increased, SADF incursions into Angola resumed, and another large-scale invasion was launched in July 1987. In November General Geldenhuys, the SADF Chief of Staff, admitted for the first time that his forces were fighting alongside UNITA. MNR activity in Mozambique intensified. Apparently still believing the ANC threat to be primarily external, the SADF struck decisively at the ANC presence in the region in 1986, most dramatically in the effective blockade of Lesotho in January which led to the coup, and in the May raids on Gaborone, Harare and Lusaka which effectively ended the Commonwealth initiative. Death and kidnap squads were deployed against the ANC even in Swaziland and post-coup Lesotho, while the October 1986 announcement of a ban on migrant labour from Mozambique was designed to force the latter to reduce the non-military presence of the ANC there.

In April 1987 an SADF raid on Gaborone killed four alleged ANC terrorists. The following month an open SADF raid on alleged bases in metropolitan Maputo signalled the demise of the last vestiges of Nkomati. Two bomb incidents in Harare, also in May, were believed to have been South African attacks directed at ANC targets. An April 1987 commando raid on Livingstone, Zambia followed only three days after Pretoria had warned the Zambian government of alleged impending ANC attacks on South African targets, to be launched from Zambian territory. In

September several parcel bombs of South African origin were discovered in Zambia.

In 1986, Pretoria became increasingly preoccupied with the threat of sanctions. It responded with economic measures designed to coerce its neighbours into opposition or at least silence on the sanctions issue. Swaziland and Lesotho, which had explicitly opposed sanctions, were exempted. Expectations that Robert Mugabe would announce the imposition of sanctions by the end of 1986 were not fulfilled, and his subsequent sanctions proposal to the Zimbabwean Cabinet in July 1987 was defeated. No other country seemed at all likely to risk sanctions, least of all Mozambique.

The first signs of a renewed phase of assertive incorporation came in January 1988, when South Africa, Portugal and Mozambique reached agreement in principle on the security and maintenance of power lines from Cahora Bassa. Later in the year Mr Botha embarked on a 'charm offensive', meeting President Chissano at Cahora Bassa where the two leaders revived the corpse of Nkomati, Mozambique in desperation hoping that this time Pretoria would keep its word. Mr Botha also visited Malawi and Zaire. But by far the most important event of 1988 was the success of the Angola-Namibia talks mentioned; the principles of a peace agreement were agreed in July, and a final agreement signed in December in what constituted a belated triumph for the Reagan administration's long discredited 'constructive engagement', albeit one made possible only by a new Soviet wish to end costly involvements in regional theatres of conflict.

The international involvement in this agreement gives reason to hope that it will, unlike Lusaka and Nkomati, be observed. There appear to be a number of reasons for this dramatic reversal of South African policy, which few would have forecast a few months earlier. The course of the war itself is one, together with renewed concern at the level of white casualties. More fundamental is Pretoria's growing realization of the unsustainable economic demands which it faced. It was simultaneously attempting to hold on to Namibia at great economic and military cost; to dominate southern Africa militarily and economically; and to maintain and develop internal security structures which effectively suppress internal opposition.

As if this were not enough in the face of a stagnant economy suffering from a net outflow of capital, the government was attempting to go much farther than ever before to meet the socio-economic needs of its black population, in the effort to legitimize new constitutional structures. Thus the 1987 budget included a 40 per cent increase in African education spending and a 26 per cent increase in the development aid budget, which funds 'homeland' development. In housing, another official priority area, the Urban Foundation estimates that R1.2 bn p.a. will be required until the end of the century to eliminate a housing backlog of 560,000 units (SAIRR, 1987: 21).

The circle cannot be squared, and some of Pretoria's commitments had to give in what appears to be the most significant fusion to date of foreign and domestic policy. The latter, and above all Pretoria's determined attempt to secure legitimacy for its constitutional restructuring, has triumphed. On a major front the 'pax Americana' has been accepted; on others the emphasis appears to have reverted, for the time being at least, to the 'pax Pretoriana'.

References

Banks, M. (1969). Systems analysis and the study of regions, *International Studies Quarterly,* **13** (4), 335–60.

Daniel, J. (1984). A comparative analysis of Lesotho and Swaziland's relations with South Africa, *South African Review*, **2**, Johannesburg, Raven Press, pp. 228–38.

Davies, R. (1987). South African regional policy post-Nkomati: May 1985 – December 1986, *South African Review*, **4**, Johannesburg, Raven Press, pp. 341–55.

Davies, R. and O'Meara, D. (1987). Total strategy in southern Africa: an analysis of South African regional policy since 1978. In I. S. R. Msabaha and T. M. Shaw (eds), *Confrontation and Liberation in Southern Africa*. Aldershot: Gower, pp. 239–78.

Geldenhuys, D. (1981). Some Strategic Implications of Regional Economic Relationships for the Republic of South Africa, *ISSUP Strategic Review*, Pretoria: University of Pretoria Press.

Green, R. *et al.* (1987). *Children on the Front Line*. UNICEF paper, cited by Hanlon (1987: 339).

Grest, J. (1987). 'Mozambique since the Nkomati Accord', *South African Review*, **4**, Johannesburg, Raven Press, pp. 356–372.

Grundy, K. W. (1973). *Confrontation and Accommodation in Southern Africa: The Limits of Independence*. Berkeley: University of California Press.

Hanlon, J. (1986). *Beggar Your Neighbours: Apartheid Power in Southern Africa*. London: Catholic Institute for International Relations and James Currey.

Hanlon, J. (1987). Relations with southern Africa: introduction, *South African Review*, **4**, Johannesburg, Raven Press, pp. 332–40.

Jaffee, G. (1983). The Southern African Development Co-ordination Conference. (SADCC), *South African Review*, **1**. Johannesburg, Raven Press, pp. 23–32.

KCA (Keesings Contemporary Archives).

Lemon, A. (1986). A geopolitical perspective. In C. A. Woodward, *On the Razors Edge: Prospects for Political Stability in Southern Africa* Pretoria: Africa Institute, pp. 19–29.

Lemon, A. (1987). *Apartheid in Transition*. Aldershot: Gower.

Lodge, T. (1983). The African National Congress, *South African Review*, 1, Johannesburg: Raven Press, pp. 50–4.

Maasdorp, G. (1980), Forms of partition. In R. I. Rotberg and J. Barratt, (eds), *Conflict and Compromise in South Africa*. Cape Town, David Philip, pp. 107–46.

Martin, R. (1988). Southern Africa: the price of Apartheid. A political risk analysis, *The Economist Intelligence Unit*, Special Report no. 1130.

Reitsma, H. J. A. (1980). Africa's landlocked countries: a study of dependency relations, *Tijdscnrift voor economische en sociale geografie*, **71**, 130–41.

SAIRR (South African Institute of Race Relations) (1987). *Social and Economic Update Two*, 2nd quarter, 1987.

Spence, J. E. (1987). Foreign policy: retreat into the laager. In Blumenfeld, J. (ed.) (1987). *South Africa in Crisis*. London: Royal Institute of International Affairs and Chatham House, pp. 76–94.

Vale, P. (1987). Regional policy: the compulsion to incorporate. In Blumenfeld, J. (ed.) (1987). *South Africa in Crisis*. London: Royal Institute of International Affairs and Chatham House, pp. 155–75.

Wallerstein, I. (1974). Dependence in an interdependent world: the limited possibilities of transformation within the capitalist world economy, *African Studies Review*, **17**, 1–26.

13 The Political Geography of Minority Control in Israel and Malaysia

Oren Yiftachel, Naomi Carmon and Dennis Rumley

Introduction

This chapter provides a step towards an understanding of the political geography of minority control in plural (multiethnic) states. In particular, it suggests a framework to examine the influence of public policy in inter-ethnic relations which, in turn, may lead to political (in)stability. The framework is used to investigate the cases of Israel and Malaysia – two biethnic democracies which have so far successfully maintained internal political stability through the control of ethnic minorities.

The chapter describes and analyses in detail one significant political-geographical element of control in the two countries: settlement strategies which in each case sought to create a high degree of spatial ethnic mix as a means of preventing minority insurgence. Finally, a proposal to study the long-term impact of these settlement policies on political stability is presented.

A theoretical framework

It is recognized from the outset that comparative cross-national analyses confront problems which arise out of the different political and geographical contexts within which the phenomena studied are located. However, cross-national studies in political geography can provide a useful method for arriving at meaningful generalizations (Verba *et al.*, 1978; Rumley, 1987). Here, a comparison is made between two biethnic democratic states which form a special type of plural democracy. These states are composed of two main non-assimilating ethnic groups which compete over political, land and economic resources. Due to common and sometimes violent interethnic tensions, governments of such states are often occupied with the task of maintaining internal political stability, defined here as a combination of social order and regime persistence (Lane and Ersson, 1987; Lijphart, 1977).

Governments in biethnic democracies have used two main approaches in attempting to maintain stability: consociation and control (see Kipnis, this volume, Chapter 15). The consociation model rests on power sharing, segmental autonomy, territorial separation and political participation (rather than pure majoritarianism) as a conflict-regulating mechanism (Lijphart, 1977). On the other hand, the control option relies heavily on the numerical strength of the dominant ethnic group typically using measures aimed at economic dependence, political containment and denial of minority territorial definition (Smooha, 1982; Lustick, 1980). Policies towards ethnic minorities in Canada, Belgium, Switzerland and pre-1963 Cyprus are examples of the consociational approach, whereas public policies in the cases of Northern Ireland, Sri Lanka and Cyprus between 1963 and 1974 constitute examples of the control option (Yiftachel, 1988).

In contrast to conventional wisdom, it has been found that the option of a liberal democracy, based on integration, individual merits and consensus-building, is not a viable alternative for plural democracies. This is articulated by Lijphart (1977: 238) who stated that: 'the real choice for plural societies is not between the British (liberal) model and the consociational model, but between consociational democracy and no democracy at all'.

A useful way to examine the consequences of public policies towards ethnic minorities is by assessing their impact on the political geography of interethnic relations. Four key components of interethnic relations are proposed here as determinants of political stability or instability. These are termed 'ethnic contrasts' and include: competition over land control, socioeconomic gaps, power disparities and political polarization (Figure 13.1). All four are – at least partially – malleable political and geographical variables, thus forming a framework within which the political-geographical impact of public policy can be evaluated.

Central to the framework is the proposition that the higher the intensity of the ethnic contrasts, the more likely they are to result in political instability. This proposition is based on the 'mobilisation of discontent' concept developed by Gurr (1970) and Gurr and Lichbach (1986) and on the inherent instability of political systems based on dominant ethnic cleavages (Horowitz, 1982; Lipset, 1959). The theoretical framework proposed here therefore distinguishes between the short- and long-term consequences of public policy. In the short-term, policies may influence the intensity of ethnic contrasts without necessarily affecting levels of political stability. However, in the long-term, persisting ethnic contrasts, reinforced by public policy, are likely to be translated into threats to the maintenance of social and political order (Figure 13.1).

Three other aspects of the proposed framework should be clarified. First, as illustrated in Figure 13.1, the 'environment' within which cases analysed are set is relevant to the consequences of public policy. Environment in this context refers to issues such as international geopolitical structures, the territorial concentration of ethnic groups, the relative size of minorities and

ENVIRONMENT

Public Policy in Biethnic Democracies	Consociation Control	Ethnic Contrasts:	Political (in) Stability:
		1. Comp. Over Land	1. Social Disorer
		2. Socioeconomic Gaps	
		3. Power Disparities	2. Threats to Regime Persistence
		4. Political Polarisation	

ENVIRONMENT

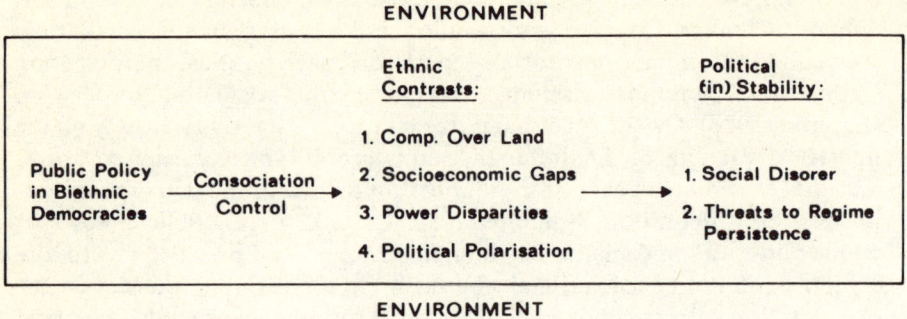

Figure 13.1 Ethnic contrasts and public policy threatening social and political order

their levels of economic, organizational and military power (Esman, 1987; Smith, 1981).

Second, the comparative scope of the proposed framework is limited. That is, it is only relevant to political systems which are formally democratic or 'open'. It has been widely accepted that minority behaviour varies markedly between democratic and non-democratic systems due to different levels of minority expectations and regime responsiveness (Gurr and Lichbach, 1986; Lijphart, 1977). The consequences of minority control in the formal democracies of Israel and Malaysia, therefore, may possess only limited relevance to the impact of control over ethnic groups under other regime types, such as in the cases of South Africa or of the Soviet Union. Finally, although the framework can be applied to the analysis of both consociational and control policies, the present chapter focuses on the latter.

Ethnic contrasts in Israel and Malaysia

Before outlining state policies towards the minority in Israel and Malaysia, it is necessary to discuss briefly the intensity of ethnic contrasts in both cases. Both the Arabs in Israel and the Chinese in Malaysia are culturally distinct, non-assimilating minorities. Arab citizens constitute 18 per cent of Israel's 4.5 million citizens in its pre-1967 borders (including East Jerusalem), while the Chinese are 34 per cent of Malaysia's 16 million people.[1] Both minorities exhibit some degree of geographical concentration: Arabs mainly live in the Galilee and the 'Triangle' regions of Israel, where they form a clear majority (Figure 13.2), while the Chinese form the majority in most of Malaysia's large cities, in the north-western state of Penang and the largest group in Selangor (Figure 13.3).

Israel is a highly centralized unitary state. Power is concentrated in government ministries and political organizations in the central cities and is seldom shared with regional or local authorities (Eisenstadt, 1985). Unlike

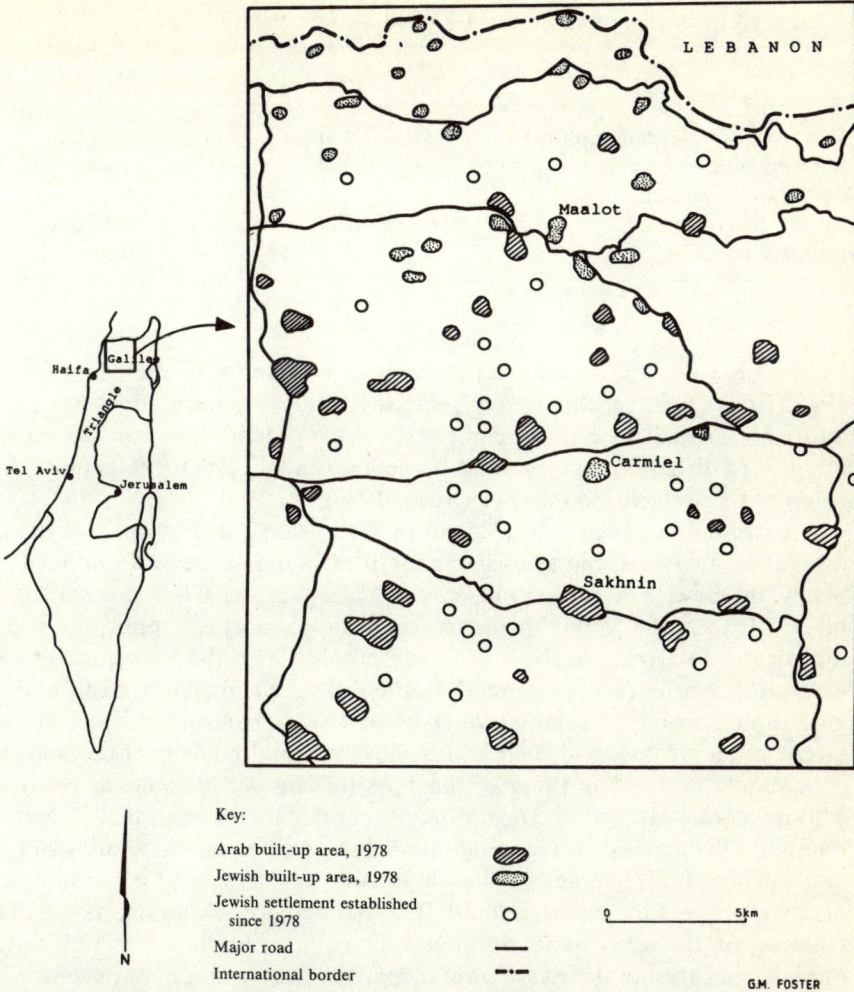

Figure 13.2 Arab and Jewish settlements in the Galilee

Israel, Malaysia possesses a federal political structure, although, in effect, political power is highly centralized. It has been suggested that the Malaysian federation was created not as a response to regional communalism, but rather as an accommodation to the historical legacy of the Malay States and to the associated institution of the Sultanate (Shafruddin, 1987). The four areas of interethnic relations listed above show pronounced minority–majority contrasts in both countries (Table 13.1). Competition over land control is intense in Israel on both natural and local levels. Israeli Arabs are part of the Palestinian people, which, like the Jewish people, claim historical rights for the land of Israel. On a local level, competition over land has been manifested in Arab struggle against land expropriation

Table 13.1   A generalized summary of ethnic contrasts in Israel and Malaysia

	Israel	Malaysia
Competition over land control	High	Medium
Socioeconomic gaps	High	High (reversed)
Power disparities	High	Medium
Political polarisation	Medium	High

by the state and the location of municipal boundaries (Soffer and Finkel, 1988; Jiryis, 1976). In Malaysia, where the Chinese minority does not claim national historical rights, competition over land control is mostly manifested in conflicts over land ownership laws, settlement policies and changes to electoral boundaries (Vasil, 1980).

Socioeconomic gaps are evident in both cases, although in different directions. In Israel, the average standard of living of the Arab minority is below the Jewish average, whereas in Malaysia, the Chinese minority is better off than the Malay majority according to most economic and social indicators. In Israel, Arabs are mostly employed in the construction and industrial sectors (21 per cent of employees in each). Arab mean annual household income of salaried workers in 1984 was 68 per cent of Jewish household's income, and their education levels and housing conditions are consistently lower than those of the Jews (al Haj and Rosenfeld, 1988). In Malaysia, the Malays are traditionally rural and occupationally tend to dominate fishing and rubber smallholding as well as public administration and defence. Their average household income in 1984 was 57 per cent of that of an average Chinese household (Government of Malaysia, 1986). The Chinese, on the other hand, dominate the commercial sector as well as the mining and forest industries. Power disparities are evident in both countries, although the gaps in Israel are more intense. In Israel, Arabs are largely excluded from positions of power in the executive, legislative and, to some degree, the judicial arms of government. In addition, Arabs are not proportionally represented in the decision-making bodies of any of the major political parties, labour unions or economic enterprises. In Malaysia, even though the Barisan Nasional coalition party embraces some non-Malays, including Chinese, access to power disproportionately favours Malays at the expense of the Chinese (Milne, 1981: 171). One of the characteristic features of Malaysian democracy has been a significant electoral bias in favour of Malays especially at the state level. This, in turn, has resulted in a disproportionally low representation of the Chinese minority in politically powerful positions, although it has been consistently represented in the national government.

Ethnic political polarization is more pronounced in Malaysia than in

Figure 13.3 The distribution of new settlements in Malaysia

Israel. Arab political organization in Israel has been constrained by internal divisions and state policies of political containment and cooptation (Mar'i, 1988; Smooha, 1982). Nonetheless, a constant rise in the vote for what can be termed 'predominantly Arab parties' can be traced over the last three decades. Support for these parties has increased from about 20 per cent in the 1950s to 59 per cent in the recent 1988 elections. In Malaysia, ethnic and political cleavages overlap significantly. Ethnicity is clearly reflected in the structure of political parties and political party support. Since independence, Malaysia has been continuously governed by Malay-dominated coalitions, with the major opposition party being the Chinese-dominated Democratic Action Party (DAP). The current Malaysian parliament is essentially split into two communal groups as a consequence of the 1986 general elections (Hanafiah, 1987). In summary, sharp ethnic contrasts exist both in Israel and Malaysia. In Israel, a strong emphasis is placed on competition over land control, while the majority group dominates both the political and the economic spheres. In Malaysia, the politicization of ethnicity is more pronounced, whereas the ethnic distribution of political power does not correspond with economic power. In both countries, the tensions deriving from these contrasts in the long-term threaten their internal political stability (Yiftachel, 1988).

Policy measures and the role of settlement strategies

The fact that a high potentiality of ethnic conflict has not caused major political instability in both Israel and Malaysia, apart from sporadic cases, can be largely explained by the effective application of control mechanisms (Esman, 1987; Smooha, 1982; Milne, 1981; Lustick, 1980). This can be favourably compared to other biethnic societies such as Northern Ireland, Sri Lanka and Cyprus where minority grievances, not effectively controlled, have caused violent interethnic conflicts and the collapse of social and political order. On a world-wide scale of ethnic violence prepared by Hewitt (1977: 152), Israel and Malaysia are both ranked significantly lower than the three cases mentioned above. In Israel, control of the Arab minority is expressed mainly in three key policy areas: economic dependence, political subordination and territorial containment (Kipnis, 1988; Smooha, 1982; Lustick, 1980). In Malaysia, following the interethnic riots of 1969, pre-existing constitutional measures favouring the Malays have been reinforced via an elaboration of the New Economic Policy. Other 'control' policies relate to the use of Malay language in schools, a national ideology, censorship, 'preventative detentions' and planned population dispersal (Esman, 1987; Hui Lim, 1985).

A significant element of the control approach adopted by the two regimes has been their territorial and settlement policies. Both governments were concerned that the creation of homogeneous territorial pockets of minority populations may result in the emergence of ethnic separatist movements (Muir and Paddison, 1981: 160). Two examples, one from each country, are discussed in order to illustrate the translation of this concern into planned settlement programmes. The analysis of the rationale, implementation methods, legal and institutional settings and consequences of these policies is central to the understanding of the political geography of minority control.

(a) The Galilee New Settlement Programme, Israel

The Galilee forms the main part of Israel's Northern District (Figure 13.2). In 1978, there were 51 Arab and 71 Jewish settlements in the region, accommodating 255,000 Arabs and 81,500 Jews (Carmon *et al.*, 1988). Most of the Arab settlements were 'villages' which had spontaneously developed into small or medium sized towns, while the Jewish settlements concentrated in five urban centres and 66 small agricultural villages. During the mid-1970s, four important political and geographical processes became evident in the hills of Galilee: (a) despite several previous Jewish settlement efforts, Arabs have remained a decisive majority in the region, forming about 75 per cent of its population; (b) a lack of rural-to-urban migration among the Arabs of the Galilee and their tendency to build low-density housing has caused extensive expansion of built-up areas in most Arab villages; (c) policy efforts to expand Jewish towns in the Galilee during the early and mid 1970s involved the expropriation of some land owned by Arabs, causing a wave of anti-government protest and violence; and, (d) several key subregions in the Galilee were left with no Jewish presence, creating difficulties in monitoring the use of state land (Carmon *et al.*, 1988; Kipnis, 1987; Soffer and Finkel, 1988).

A strongly held belief among Israeli (Jewish) policy-makers during this period was that a real danger existed of 'losing' the Galilee to the Arab minority. It was frequently stated in policy documents that the Arabs in the region 'invade' state land for individual and nationalistic reasons, and that this phenomenon should be controlled (for example, see the Jewish Agency, 1978, 1979 a–c, 1980; Jewish National Fund, 1979; Israeli Government, 1979; Survey, 1988).[2] The fear of losing the Galilee can also be related to the persisting broader Israeli–Palestinian conflict and, in particular, to the 1947 UN partition proposal for Palestine, under which the area in question was designated for an Arab-Palestinian state.

The main goals of the Galilee new settlement programme were shaped by the circumstances mentioned above. Although no single document was ever produced as the '1975 New Development Plan' and no clear set of goals was ever explicitly articulated, the documents reviewed revealed three main goals: the protection of state land, the dispersal of Jews throughout all parts of the Galilee, and an increase in the overall proportion of Jews in the region. The emphasis was clearly placed on the first two goals, as evident from the following statements: 'the plan is aimed at securing (Jewish) control over state land in regions where Jewish population is scarce . . . Jewish presence is essential for the protection of land' (Jewish Agency, 1979a).

To fulfil these goals, the first stages of the new settlement programme called for the rapid establishment of '*mitzpim* ' – 'lookout' posts. These settlements were designed as temporary camps of several households, around which a permanent settlement would eventually develop. Twenty-nine such *mitzpim* were established during 1979 and 1980. Thirty additional (permanent) settlements were added during the second stage which lasted from 1980 to 1985. They were located in key strategic positions either on hill tops (for observation purposes) or as wedges between expanding Arab villages (Figure 13.2). The new settlers were of relatively high socioeconomic status, being mainly attracted to high living standards and clean environment offered in the Galilee. At the end of 1988, the average size of the new settlements reached 35 households, the smallest with seven and the largest with over 300 (Carmon *et al.*, 1988).

The geopolitical strategy adopted by Israeli planning authorities in the Galilee was similar to the approach used in the Jewish settlement of the West Bank. Over 100 small-medium Jewish settlements have been established since the mid-1970s, many of them in areas heavily populated by Arabs (Newman, 1984). This concerted attempt to increase the spatial mix between Arabs and Jews in all parts of the Land of Israel (Palestine) was closely linked to the rise of the right-wing Likud to power in 1977.

However, while the issue of Jewish settlement in the occupied West Bank has been controversial, attempts to increase the Jewish population of the Galilee have enjoyed consensus among the Jewish public. This was illustrated by the release of the 'Judaisation of Galilee' strategy by the Labour-led Israeli government in 1975 (Israeli Government, 1975). This strategy committed the government to the creation of housing, jobs and infrastructure, mainly in Galilee towns, to enable the absorption of new Jewish settlers. The new settlement strategy of the late 1970s thus built upon the widely accepted foundations of earlier Israeli plans to attract Jews to the Galilee.

Importantly, the main planning and implementation body of the new settlement strategy was not the Israeli government but the Jewish Agency and, to a lesser degree, the Jewish National Fund. Both these bodies have been previously described as 'quasi-governmental' (Yiftachel, 1987), due to their legal 'covenants' with the Israeli government. This operational structure has caused difficulties among the Arab minority, as these two quasi-governmental agencies are not accountable to the general Israeli public, but to the Jewish diasporas. Despite this, they hold significant land use planning powers, enshrined in the above-mentioned covenants, mainly pertaining to the establishment of new settlements, roads and forests (see also Vilkanski *et al.*, 1988). In the case of the Galilee settlement plan, the Israeli government approved a broad brush plan (Israeli Government, 1975, 1979), leaving the details to be worked out by the two bodies. The Israeli government itself, however, has also been directly involved in most stages of the programme. In particular, it built a large part of the housing and infrastructure through the country building branch of the Ministry of Housing and Construction, and devised a 'fast-track' method to approve the statutory outline plans of the new settlements.

The last point can serve to illustrate how the circumstances in the Galilee, described earlier, affected the planning process. Usually, the approval of statutory land use plans (outline plans) in Israel is very slow. The new settlement plan was proposed and implemented with a sense of emergency and urgency. For this reason, the Northern District Committee for Planning and Building, under the auspices of the Minister of the Interior, approved the plan in an unusual manner. The committee used the 1946 statutory District Plan for the Galilee (produced during the British Mandate) which legally was still in force but had not been used since 1975. The 1946 plan allowed the committee to approve the new settlements directly, thus shortening the legal procedure considerably and by-passing the requirement for public comment on the plan included in the later Israeli Planning and Building Law (1965).

Overall, the Galilee new settlement plan can be seen as an integral part of the policy of control over the Arab minority in Israel. The control elements of the strategy appear both in its goals and in its means of implementation. The goals of stopping the Arabs from using state land and handing much of the land in question to Jewish settlements and the positioning of Jewish settlements so as to overlook Arab villages were clearly designed to control a strengthening ethnic minority. The goals of limiting the expansion of Arab villages and establishing wedges between concentrations of Arab population were also aimed to reduce the territorial base and therefore the political power of the minority. The rapid manner in which the plan was implemented was, at least in part, aimed at surprising the local Arab population, thus preventing it from initiating a public campaign against it.

It must also be mentioned that the demographic impact of the strategy was small (about 4,800 people by November 1988), and that some economic benefits such as the use of newly constructed roads to increase employment opportunities, flowed to the Arab population (Survey, 1988). However, the new settlement plan added a visible and physical dimension to the control of the Arab minority in Israel, which as mentioned, was perceived by Jewish decision-makers to be potentially hostile to the state (Soffer and Finkel, 1988; Carmon *et al.*, 1988, Survey, 1988).

(b) The 'new village' programme in Malaysia

In Malaysia, the 'new village' programme was an ethnically oriented, politically based 'resettlement' policy which has had an enduring impact upon that state's settlement pattern and upon the population of the Chinese community (Figure 13.3). The programme has been described as a massive exercise in social engineering which involved the relocation of about 25 per cent of all Malaysian Chinese (Strauch, 1981: 63). Between 1947 and 1957, for example, it had the effect of increasing the number of urban places (settlements with more than 1000 inhabitants) by about 150 per cent (Sandhu, 1964: 177). The policy also had considerable social and economic effects. The Malaysian 'landlord class' for example, came to replace the former Chinese retailers as a consequence of relocation (Gullick and Gale, 1986: 244). It has been estimated that 445 'new villages' had forcibly been created by 1955 containing approximately 550,000 people (Voon and Khoo, 1986: 37).

Until the 1930s, the vast majority of the Chinese community in Malaysia was concentrated in cities and mining areas, but even before the Depression there was some dispersion of population to subsistence or 'squatter' farms along the fringes of the jungle. One of the main reasons for such squatting was that all unalienated land was vested in Malay rulers, much of which in turn could only be alienated to Malays. Restrictions on the acquisition of land were thus differentially applied to ethnic groups, especially the Chinese (Sandhu, 1964: 157).

With the Japanese occupation, the rate of rural dispersion markedly increased, partly in response to the overall shortage of food, but also in response to Chinese concerns over possible Japanese 'vindictive animosity'. The squatters became 'the Ishmaels of Malaya against whom every man's hand was raised' (Gullick and Gale, 1986: 87). By 1945, there were approximately 400,000 dispersed Chinese 'squatters' in Malaysia. During the occupation, the squatters became one of the major sources of resistance and provided recruits to the communist-led Malayan People's Anti-Japanese Army (MPAJA). After World War II, the MPAJA regrouped as the Malayan Communist Party and initiated a protracted jungle war of independence known officially as the 'Emergency'. In essence, the jungle was the guerrilla base with food and support being supplied from the squatter settlements.

From a strategic viewpoint, the British authorities at the time felt that if the lines of support to the communist guerrillas could be cut, then the insurgency might be overcome. Initially, the government policy was to round up squatter settlement sympathizers and repatriate them, mainly to China. Approximately 25,000 squatters were dealt with in this manner, but it was clear that this policy would be unwieldy given the large numbers involved (Gullick and Gale, 1986: 91). In 1950, following federal-state conflict over the implementation of squatter policy, the so-called Briggs plan was enacted and was designed to 'resettle' the squatters forcibly into 'new villages'. Although the 'new village' policy has been seen primarily as a military matter (Strauch, 1981: 60–3), from a political-geographical perspective its main goal was the spatial identification, containment and separation of the inhabitants in order to prevent them being either victims or supporters of the guerrillas (Voon and Khoo, 1986: 38). In addition, it has also been argued that the new villages plan was another example of policies of coercion being used on those members of Malaysian society whose complete loyalty was perceived by the authorities to be in some way doubtful (Strauch, 1981: 40).

It has been noted that, to a degree, the term 'new village' is a misnomer since not all of the location sites were in fact new. Three types of new village have been identified – those which were entirely new; those built around and absorbing small existing settlements; and those established as additions to larger settlements (Sandhu, 1964: 168–70). An important criterion for the location of new villages was that they should be within a reasonable distance from original workplaces. More than 200 villages were thus located in controlled sites near rubber holdings or mines. In addition, a further 267 new villages were built around or attached to existing settlements (Voon and Khoo, 1986: 38–41). Other locational criteria included defensibility and agricultural and economic potential (Gullick and Gale, 1986: 92).

In essence, the method of relocation involved the forcible uprooting of half a million farmers who lived on the jungle fringes and relocating them inside barbed-wire-enclosed new villages which at least initially were more like large prison camps where movement was strictly controlled (Kim Hoong, 1987: 21). In most cases squatters were resettled within 3 to 8 kilometres from their original location, although in some cases moves of longer distance were required. The largest number of people (32%) and villages (30%) were relocated in Perak (Figure 13.3), while significant relocations also occurred in Johore (25% people and 22% villages) and in Selangor (18% people and 9% villages) (Voon and Khoo, 1986: 39–42). Perhaps surprisingly, little violence or physical resistance occurred on the part of the Chinese during the actual relocation process (Siaw, 1983: 99–100).

The enforced resettlement programme associated with the 'squatter problem' is associated with some important legal implications. For example, from a constitutional viewpoint, Articles 89 and 90 provide for Malay privileges in relation to the reservation of land under which the government may acquire land for Malay purposes (Groves, 1980: 23). The constitution thus serves to discriminate against the Chinese community in terms of land ownership thereby encouraging their urbanization. The new village programme itself clearly has been part of the urbanization process of the Chinese community. The process of Chinese urbanization in Malaysia in turn reduces their opportunity to participate electorally given the operation of Malaysian electoral laws and procedures in terms of malapportionment and electoral redistricting.

Given the constraints on Chinese land ownership, many of the squatters illegally occupied and cultivated State land. Some squatters also occupied State land under a Temporary Occupation Licence (TOL) on an annual basis giving illegal occupation a quasi-legal status (Voon and Khoo, 1986: 36). Other squatters occupied Malay reservations again illegally from a constitutional viewpoint (Siaw, 1983: 80).

Clearly, the issue of land ownership was critical to the likely viability of the newly created villages. Although new villagers were encouraged to apply for title to their house lots, many were reluctant and the time required for the successful allocation of legal title varied considerably by State (Voon and Khoo, 1986: 41). Given that the new village programme was carried out during the turmoil of the Emergency and that the whole operation which involved so many people was completed very quickly, it is not surprising that there were some technical, financial and legal problems. However, apart from a group of new villages in Kelantan, most settlements experienced a large initial increase in permanent population (Nyce, 1973: 192–3). By the end of the Emergency, the barbed wire fences had been removed and by the early 1960s many new villages had developed into prosperous settlements (Sandhu, 1964: 180). Current planning for the new villages is complex and appears to occur at different levels. On the one hand, there is still no overall machinery to administer new villages and finance village development (Voon and Khoo, 1986: 52). However, to a certain extent the new villages have been incorporated into Malaysian development programmes. By 1979, conditions had improved and residents were involved in a variety of occupations including smallholdings, farm and estate labouring, and some were self-employed service workers in nearby urban centres. From the outset, however, the planning and provision of essential services has been problematic. For example, in 1979 only about 50 per cent of villages had access to potable water and about two-thirds were equipped with health facilities (Government of Malaysia, 1979: 33). By 1985, on the other hand, most of the new villages had proper water, electricity and schools. In addition, the total new village population had reached about 1.8 million (Government of Malaysia, 1986: 91–6).

At a second level, some new villages are necessarily incorporated into the structure plans of major urban centres. However, in the cases of Kuala Lumpur and Ipoh, for example, the new villages do not appear to have been given due attention via a variant of spatial discrimination. This problem relates to the overall problem of a lack of local control – that is, most new villages are incorporated or absorbed into a town, district or urban authority. A lack of recognition and articulation of local planning issues coupled with the problem of competing claims upon scarce resources within an encapsulating political entity, can often be to the disadvantage of the new village (Voon and Khoo, 1986: 50–2).

Overall, five main elements of control were embedded in the Malaysian new village programme. First, the programme caused the relocation of a large percentage of the Chinese community. Associated with this was a policy of spatial identification and containment of potential communist sympathizers. This undoubtedly had a decisive impact upon communist insurgents (Sandhu, 1964: 180). Second, it created constitutional discrimination in favour of Malays over land ownership. Problems of lack of security over land tenure within the new villages remain (Voon and Khoo, 1986: 46). Third, the containment of Chinese in the new villages has had the effect of reducing their opportunities for meaningful political participation given the operation of Malaysian electoral procedures. Fourth, the administrative encapsulation of new villages minimized local political autonomy and the articulation of legitimate local needs. Finally, control is reinforced in the new villages and order and stability are maintained by the police, who are seen as 'Malay outposts' in areas inhabited primarily by Chinese (Zakaria, 1987: 121).

A concluding remark

In overview, it is illuminating to note that a similar political-geographical goal of creating a spatial mix of majority and minority groups has guided policy-makers in both Israel and Malaysia. In both cases, the settlement strategy has been a physical and visible dimension of control and has had a significant impact upon the distribution of ethnic populations. The methods of implementation, however, have been significantly different: in Israel, majority groups have penetrated minority areas, whereas in Malaysia, minority groups have been resettled into areas controlled by the majority. Another important difference exists between the two cases in terms of the method of implementation: the Israeli government used incentives to attract voluntary population to settle in the new settlements, while the Malaysian government used its coercive power to implement the new village programme. Both settlement strategies achieved their short-term goals of preventing the minority from developing an ethnically homogeneous territorial concentration which could have destabilized the political systems of the two countries. These strategies thus provide important examples of how regimes in biethnic societies are capable, through the implementation of population dispersal strategies, of maintaining the domination of the larger ethnic group and preventing the territorial expression of minority grievances. Most importantly, however, the long-term consequences of these policies in terms of their impact on interethnic relations and political stability are less clear. A meaningful evaluation of such consequences may need to be conducted decades after implementation. Such evaluation necessitates a detailed examination of the influence of the control policy on all four areas of ethnic contrasts suggested by the framework formulated earlier (Figure 13.1): competition over land control, socioeconomic gaps, power disparities and ethnic political polarization. Research within this framework is currently in progress on the Galilee region (Yiftachel, forthcoming), forming a potential point of departure for future research on the subject both in Israel and other comparable contexts. Future comparative and case-study research is required in order to elaborate more fully the various dimensions of the political geography of control.

Notes

1. Due to their relatively small size, the paper excludes the Indian and other small ethnic groups in Malaysia.

2. The survey included 44 interviews which were conducted with Israeli decision-makers (particularly in the land-use planning and settlement policy areas) and Arab local and national leaders. Israeli decision-makers included past and present Commissioners of the Northern District in the Ministry of Interior, past and present Chief planners of the Northern District, office bearers in the Israeli Land Authority, the Ministry of Housing and Construction, the Jewish Agency, the Jewish National Fund, professional participants in the preparation of plans for the Galilee and planners of individual Arab settlements. Arab interviewees included 16 heads of Arab local councils in the Galilee, past and present, Chairmen and secretaries of the Arab National Committee of heads of Local Councils, Chairman of the National Committee for the Defence of Arab Lands, and past and present members of the above two committees. Due to the request of most interviewees, their names are kept with the authors.

Acknowledgements

The authors wish to thank the Albert Einstein Research Fund (through the Technion) and the University of Western Australia for their financial support which made this research possible.

References

al-Haj, M. and Rosenfeld, H. (1988), *Arab Local Government in Israel*. Tel-Aviv, The International Centre for Peace in the Middle East.

Carmon, N., Czamanski, D., Kipnis, B., Law Yone, H. and Lifshitz, G. (1988), *New Settlement in the Galilee – An Evaluation*. Haifa, Centre for Urban and Regional Research, the Technion (Hebrew).

Eisenstadt, S. N. (1985). *The Transformation of Israeli Society*. London: Weidenfeld & Nicolson.

Esman, M. J., (1987), Ethnic politics and economic power. *Comparative Politics*, **19**, 395–418.

Government of Israel (1975), *The New Development Plan for the Galilee*. Jerusalem: Government Information Centre (Hebrew).

Government of Israel (1979), *Decisions of the Settlement Committee, Government Secretariat*. Jerusalem, unpublished (Hebrew).

Government of Malaysia (1979). *Mid-Term Review of the Third Malaysian Plan*. Kuala Lumpur: Government of Malaysia.

Government of Malaysia (1986), *Fifth Malaysia Plan (5MP), 1986–1990*. Kuala Lumpur.

Groves, H. E. (1980), Malaysia. Volume 9. In A. P. Blausten and G. H. Flann (eds.), *Constitutions of the Countries of the World*. New York: Oceana.

Gullick, B. and Gale, B. (1986), *Economic Development in Malaysia*. Petaling Jaya: Pelanduk.

Gurr, T. (1970). *Why Men Rebel*. New Jersey: Princeton University Press.

Gurr, T. and Lichbach, M. (1986), Forecasting internal conflict. *Comparative Political Studies*, **9**, 3–38.

Hanafiah, A. M. (1987), The Malaysian general election of 1986. *Electoral Studies*, **6**, 279–85.

Hewitt, C., (1977), Majorities and minorities: a comparative survey of ethnic violence. *Annals of the American Academy of Political and Social Sciences,* **433**, 150–60.

Horowitz, D. (1982), Dual authority polities. *Comparative Politics*, **14**, 329–50.

Hui Lim, M. (1985), Affirmative action, ethnicity and the integration: the case of Malaysia. *Ethnic and Racial Studies*, **8**, 250–76.

Interviews (1988) conducted by Oren Yiftachel during 1988 in Israel.

Jewish Agency, Settlement Department (1978), *National Land in the Galilee*. Safed: Galilee District, Mimeo, Unpublished (Hebrew).

Jewish Agency, Settlement Department (1979 a, b, c), *Working Papers on the Mitzpim Plan, Galilee and Northern Districts*. Haifa and Safed, mimeo, unpublished (Hebrew).

Jewish Agency, Settlement Department (1980), *A New Settlement Project for Nahal Tzalmon*. Safed: Galilee District (Hebrew).

Jewish National Fund (1979), *Land Use Plan for the Galilee*. Safed, Galilee and Northern Districts (Hebrew).

Jiryis, S. (1976), *The Arabs in Israel*. Jerusalem: Monthly Review.

Kam Hing, L. (1987), Three approaches in peninsular Malaysian Chinese politics: the MCA, DAP and Gerakan. In H. A. Zakaria (ed.), *Government and Politics of Malaysia*. Singapore: Oxford University Press, pp. 71–93.

Kim Hoong, K. (1987), The early political movement before independence. In H. A. Zakaria (ed.), *Government and Politics of Malaysia*. Singapore: Oxford University Press, pp. 11–37.

Kipnis, B. (1987), Geopolitical ideologies and regional strategies in Israel. *Tijdβchrift voor Economische en Sociale Geografie*, **78**, 125–38.

Kipnis, B. (1988), Regional development and strategy considerations in multi-community land of Israel. In Hofman, J. (ed.), *Arab-Jewish Relations in Israel: A Quest in Human Understanding*. Bristol, Indiana: Whyndham Press, pp. 21–44.

Lane, J. E. and Ersson, S. O. (1987), *Politics and Society in Western Europe*. London: Oxford University Press.

Lijphart, A. (1977), *Democracy in Plural Societies: A Comparative Exploration*. New Haven: Yale University Press.

Lipset, S. M. (1959), *Political Man*. New York: Garden City.

Lustick, I. (1980), *Arabs in the Jewish State: Israel's Control over a National Minority*. Austin: University of Texas Press.

Mar'i, S. (1988), Sources of conflict in Arab-Jewish relations in Israel. In Hofman, J. (ed.), *Arab-Jewish Relations in Israel: A Quest in Human Understanding*. Bristol, Indiana: Whyndham Press, pp. 1–21.

Milne, R. S. (1981), *Politics in Ethnically Bipolar States*. Vancouver: UBC Press.

Muir, R. and Paddison, R. (1981), *Politics, Geography and Behaviour*. London: Methuen.

Newman, D. (1984), Ideological and political influences on Israeli urban colonisation: the West Bank and the Galilee Mountains, *Canadian Geographer*, **28**, 143–55.

Nyce, R. (1973), *Chinese New Villages in Malaya: A Community Study*. Kuala Lumpur: Malaysian Sociological Research Institute.

Rumley, D. (1987), Structural effects in different contexts. *Political Geography Quarterly*, **6**, 31–7.

Sandhu, K. (1964), Emergency resettlement in Malaysia. *Journal of Tropical Geography*, 157–83.

Shafruddin, B. H. (1987), *The Federal Factor in the Government and Politics of Peninsular Malaysia*. Singapore: Oxford University Press.

Siaw, L. K. L. (1983), *Chinese Society in Rural Malaysia*. Kuala Lumpur: Oxford University Press.

Smith, A. D. (1981), *The Ethnic Revival*. Cambridge: Cambridge University Press.

Smooha, S. (1982), Existing and alternative policy towards the Arabs in Israel. *Ethnic and Racial Studies*, **5**, 71–98.

Smooha, S. (1987), Jewish and Arab ethnocentrism in Israel. *Ethnic and Racial Studies*, **10**, 1–26.

Snodgrass, D. R. (1980), *Inequality and Economic Development in Malaysia*. Kuala Lumpur: Oxford University Press.

Soffer, A. and Finkel, R. (1988), The *Mitzpim* in the Galilee – first conclusions. In Schwartz, D. and R. Bar-El (eds), *Issues in Regional Development*. The Israeli Institute for Regional Science and the Urban and Regional Settlement Research Centre, Jerusalem: Achava Press, pp. 63–88 (Hebrew).

Strauch, J. (1981), *Chinese Village Politics in the Malaysian State*. Cambridge: Harvard University Press.

Trindade, F. A. and Lee, H. P. (1986), *The Constitution of Malaysia: Further Perspectives and Developments*. Singapore: Oxford University Press.

Vasil, R. K. (1980), *Ethnic Politics in Malaysia*. New Delhi: Radiant.

Verba, S., Nie, N. H. and Kim, J. O. (1978), *Participation and Political equality: A Seven-Nation Comparison*. Cambridge: Cambridge University Press.

Vilkansky, R., Law Yone, H. and Meyer-Brodnitz, M. (1988), Evaluating the influence of regional plans: the Galilee case study. In Schwartz, D. and Bar-El, R. (eds), *Issues in Regional Development*. Jerusalem, The Israeli Institute of Regional Science and the Centre for Urban and Regional Settlement Research, pp. 89–106 (Hebrew).

Voon, P. K. and Khoo, S. H. (1986), The new villages in peninsular Malaysia: a socioeconomic perspective. *Malaysian Journal of Tropical Geography*, **14**, 35–55.

Yiftachel, O. (1987), Urban and regional planning in Israel: structure, problems and achievements. *Australian Urban Studies*, **15**, 8–13.

Yiftachel, O. (1988), Geopolitical aspects of stability in a biethnic democracy: the case of Israel's policy towards its Arab minority. *Politics*, **23** (2), 48–56.

Yiftachel, O. (forthcoming), *State Policies and Political Stability in Biethnic States*. Ph.D. Thesis, University of Western Australia and Technion – Israel Institute of Technology.

Zakaria, H. A. (1987), The police and political development in Malaysia. In Zakaria, H. A. (ed.), *Government and Politics of Malaysia*. Singapore: Oxford University Press, pp. 111–27.

Zakaria, H. A. (1989), Malaysia: Quasi-democracy in a divided society. In Diamond, L., Linz, J. J. and Lipset, S. M. (eds), *Democracy in Developing Countries, Volume 3, Asia*. Boulder: Rienner, pp. 347–381.

14 Overcoming the Psychological Barrier: The Role of Images in War and Peace

David Newman

I tried to ask him where he got his facts from, but he was moving on, summing up fresh ammunition. After all, what people imagine can be more damaging than the truth (Hunter Davies, *A Walk along the Wall*, 1984: 163)

The cognitive environment and the geography of propaganda

Images play a major role in the development of human attitudes. The importance of perception studies, as they are reflected in mental images and selective spatial awareness, has received much attention in the behavioural sciences in general, and geography in particular (Gould and White, 1974; Downs and Stea, 1977; Pocock and Hudson, 1979). The political socialization process, through which people are influenced by their immediate environment, the education they receive and the media images they absorb, clearly influence the formulation of their own personal attitudes to political, social and economic issues. Boulding (Chapter 10, this volume) differentiates between 'folk' and 'scholarly' images, the former being attained in ordinary daily life, the latter through formal education, particularly schools and universities. The channels of communication through which images are transmitted vary from school to media and from advertisments to blatant political propaganda.

In turn, each of these influences may be manipulated (both consciously and subconsciously) by the environmental, education and media managers who wish to disseminate a specific message. Hall (1981) has used the term 'propaganda geography' to describe these influences, noting four agents of political socialization: graphics, numerics, semantics and technologics as methods through which propaganda pervades all aspects of life.

Relph (1976) has categorized place experiences as a spectrum ranging from the deepest authentic relationship to the most inauthentic, superficial experiences provided by the mass media. Instead of authentic place experience which arises from an ongoing communion between a person and his environment, the use of propaganda images brings inauthentic experiences to the fore. The media and other channels of mass communication are important agents of political and territorial socialization, through both their explicit and implicit use of graphics and semantics for the dissemination of propaganda (Ager, 1978; Burgess and Gold, 1985; Burnett, 1985; Hall, 1981;

Moore and Golledge, 1976; Muir and Paddison, 1981). The media thus 'moulds individual and social experiences of the world and in shaping the relationship between people and place' (Burgess and Gold, 1985: 1).

The use of images plays a crucial role in the development of national stereotypes. 'Other' nations and ethnic groups are characterized by specific behavioural attributes, often to their detriment. Boulding (Chapter 10, this book) notes that all images of the nation state tend to be a distortion of reality, and that scholarly knowledge is perverted by the national state in that it presents the nation state as being at the centre of history and of the world. Military images of the nation state tend to be even more prominent since their very existence is dependent on there being a potential or actual enemy, itself an image which strengthens national differentiation. In conflicts between nations, images are exaggerated in order to portray the enemy negatively, thus arousing a continued sense of antipathy towards the foe.

Conflictual images may be temporal or spatial. Temporal images involve a singular interpretation of history, focusing on past injustices, so as to provide self-justification for the ongoing conflict and the suffering it may cause. In this respect, the conscious emphasis on national awareness and feelings of national discrimination vis-à-vis other national groups, provides a powerful self-legitimation for violent conflict. Spatial images may also be used in situations of territorial and national conflict, to delineate past or present national territory or boundaries.

The process of environmental learning and resultant spatial and territorial images which make up their cognitive environments is of central concern to understanding of the conflict between Israelis and Palestinians. Waterman (1979), following Wright, has already noted the importance of the 'illusory subjective' and the 'promotional imagination' as tools used by Israelis in converting *terrae incognitae* into *terrae cognitae* (Waterman, 1979: 178-9). This chapter concentrates on the mutual spatial and territorial graphics and semantics of the two population groups. The resulting images influence both youth and adults alike in the development of mutually antagonistic stereotypes and claims.

Images in the Arab-Israel conflict: attitudes, awareness and ignorance

The formation of mental images and general attitudes does not take place in a vacuum. Israelis live in constant tension, with crises peaking reached at times of war or following terrorist attacks on civilian targets. The Arabs are perceived as constituting a collective enemy whose objective is to destroy the State of Israel and to drive its inhabitants into the sea. By extension, suspicions are also directed towards Israeli Arabs who are full citizens of the State. The recent radicalization of Israeli Arabs – particularly following the onset of the Intifada in 1987 – have only served to strengthen further these feelings on the part of Israeli Jews. A recent study by Arian *et al.* (1988) shows that while the majority of Israelis feel confident that the Israeli army will be able to cope with any military threat from the surrounding Arab nations, there nevertheless remains a deep and persistent feeling of being threatened and under siege. These attitudes comprise a 'religion of security' which 'binds the nation together despite differences of partisan politics' (p. 51) and

> Just as a child accepts unquestioningly (at least for an initial period) the religion he was born into and some of the basic answers he receives from his parents and other socializing agents regarding the mystery of creation and the existence or non-existence of the deity, so too does the Israeli child absorb at a very early age the basics of the core-belief of national security (p. 83).

Much of the mutual attitudes and awareness held by each side for the other stems from ignorance. Jews and Arabs in Israel are highly segregated – all villages are mono-ethnic, while a population mix is to be found in only a few urban centres and even then in separate neighbourhoods. Children attend either Jewish or Arab schools and in many cases do not come into any meaningful contact with the 'other' group throughout their childhood and adolescence. Attempts to bring about minimal contacts between Arab and Jewish high school students have met with only limited success. Many parents have refused to let their children attend such events. While Arabic is taught in some Jewish schools, most Israelis leave school without even a smattering of the language. Most Israeli Jews do not understand Arabic and can communicate with Arabs only in Hebrew. As a result, many Israeli Arabs and Palestinians in the West Bank understand or speak Hebrew, often because of their dependence on this language for employment or other official purposes. While the knowledge of the 'other' language arises out of necessity, few, if any, of the conflict participants read the popular literature or the media of the 'other'. Ignorance leads to estrangement which, in turn, brings mutual antagonism to the fore.

School systems are separate between and within the national groups. The political and territorial socialization processes in Arab schools are different to those in the Jewish schools, while within the Jewish sector, religious schools disseminate different political messages to those of their secular counterparts.

In the Jewish sector there is not necessarily a standardized interpretation of political events, past or present. The national religious and right-wing images are relatively closed, unwilling to assimilate changing messages. As an example, they relate to the 'cold' peace with Egypt rather than bring out the obvious benefits of the existence of such an agreement. The 'Peace Now' images are more open in that they are relatively flexible and can adjust to new information (Muir and Paddison, 1981: 46; Newman and Hermann, forthcoming). Others, for their part, view such 'open' images as suicidal and argue that any form of flexibility – in terms of holding negotiations with the PLO, or withdrawing from some of the occupied territories – threaten the continued existence of the State. It is these latter groups which have managed successfully to project themselves as constituting the 'national camp', implying that alternative views are contrary to the national interest.

In the Israel–Palestinian context, Romann (1984), Spielmann (1986) and Portugali and Newman (1987) have all made reference to the images held by one population group for the other. In Romann's (1984) study of Jerusalem, the selective awareness held by Jewish and Arab schoolchildren in the city for areas belonging to the 'other side' is studied. Despite residing in a functionally united city, Arab and Jewish residents lack basic locational knowledge concerning sites in the 'other side', even among those who have grown up since 1967 and have only ever known a unified city. Visits by Jews or Arabs to the opposite sector only take place in order to satisfy demands which cannot be met within their own sector. Owing to the dominant-subordinate system of relations, this results in many more Arabs being obliged to visit the Jewish sector (to find employment or to visit government offices) than vice-versa, these visits often being accompanied (on both sides) by feelings of discomfort.

Interaction between West Bank Palestinians and Jewish settlers is also minimal. In their study of settlers and migrant workers, Portugali and Newman (1987) note the selective spatial awareness held by residents of one settlement for their ethnically-different neighbours. Respondents were asked to name the ten nearest settlements to their own place of residence. With few exceptions (these usually being the case of large towns) residents only identified settlements which belonged to their own ethnic sector, regardless of whether there were other settlements inhabited by the other ethnic group and which were located much closer. Each of the two population groups has developed its own spatial images of the region in which they reside, these images largely ignoring the existence of the settlement system of the other group. Consumption patterns also relate to this sectoral spatial system. Each group satisfies most of its needs in the ethnic group's own settlements, regardless of whether these are the closest or most conveniently located.

The use of graphics

The use of maps by skilled manipulators plays an important part in the propaganda process (Ager, 1978; Burnett, 1985). Within the Israel-Palestinian context, the use of imagery, both graphic and semantic, is highly developed. Both national groups use maps to define the territory they claim as belonging to them. Since the very essence of the conflict between Zionist and Palestinian nationalist movements is based on the mutual claim to the same piece of territory, the use of maps is of major significance in moulding the collective claim. Benvenisti (1987) describes the reasons behind the map-making process as follows:

Coming to a new society is in a way like creation, and your first impulse is to impose on the new landscape some sort of order. Map drawing and naming of physical features is an act of possession, of creating a new reality . . . Map makers forget that their productions are merely approximations, they begin to believe in the ultimate truth of their representation . . . All immigrant societies drew maps and renamed places, because then the geography became their own. They framed the landscape with names in their own language that had significance for them (Benvenisti, 1987: 192–3).

Historical maps, depicting the ancient Biblical Jewish kingdoms, are used in order to strengthen the link between man and land. The national territory under dispute is 'Palestine' to the Palestinians and 'Israel' to the Israelis. It has largely been forgotten that until 1948, the Zionist movement desired an independent State in *Palestine*, while it is less than 20 years since the fund raising *Joint Palestine Appeal* changed its name to the *Joint Israel Appeal*. The respective images conjured up by the term Palestine are therefore time-specific (Figs. 14.1 and 14.2).

In some religious schools in Israel, Jewish children are taught that the 'true' boundaries of a sovereign kingdom are those of the 'Land of Israel' as defined in the Bible, stretching far beyond Israel's present boundaries to the south, north and east (Fig. 14.2). The refugee problem is often dealt with by the presentation of maps depicting Israel's small, densely populated territory in relation to the vast territories of the surrounding Arab countries, much of which are unsettled. In Palestinian schools, the border between (pre-1967) Israel and the Arab states does not exist, the whole territory bearing the name Palestine. Images of an Israel, its size blown up disproportionately, portray the picture of a country which threatens the whole region (Fig. 14.3).

The 'Green Line', marking the boundary separating Israel from the West Bank between 1948 and 1967, remains strongly imprinted in the minds of both Palestinians and Israelis, despite the fact that it has not existed for 23 years. Even those born after 1967 and having no other spatial memories apart from that existing today, are conscious of their crossing from Israel into the West Bank (or vice-versa) despite the absence of a visible boundary. Given the conditions of extreme conflict which govern the behaviour of both Israelis and Palestinians, the image is often stronger than the physical reality. But it is the image which more correctly mirrors the social and political realities, despite the physical forms of interaction and territorial integration.

Use is also made of *strategic*, or functional, maps. These emphasize the 'small' size of Jewish Israel in relation to the 'vast' extent of the Arab Middle East (Fig. 14.4). At a more local scale, the pre-1967 'green line' boundary is redrawn on the map of Israel, pictures of artillery are placed on the Jordanian and Syrian borders directed towards the densely populated Israeli metropolitan centres, and distances from the borders to these centres are emphasized in bold characters (Fig. 14.5). When faced with these

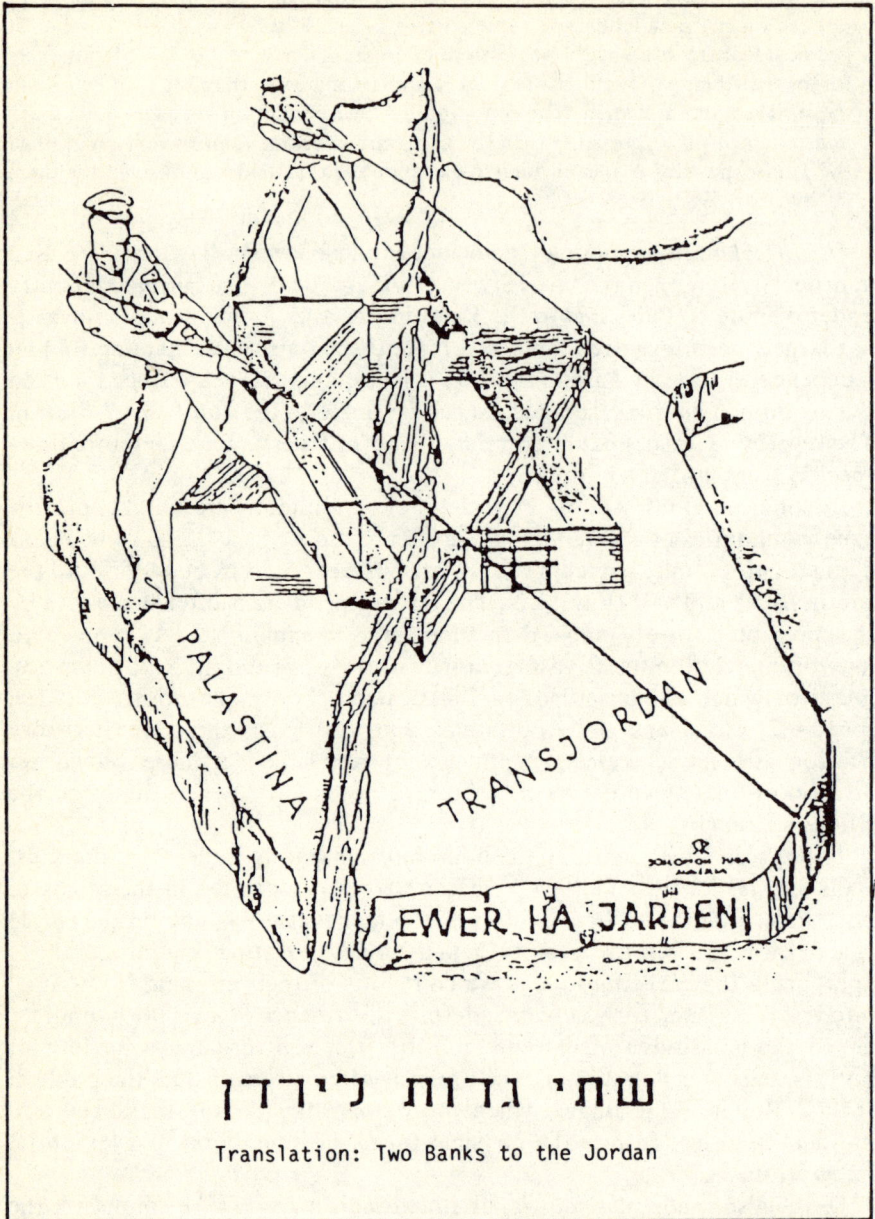

Figure 14.1 A view of Palestine (pre-1948) of the Bethar Youth Movement in Austria.

The diagram depicts the 'wholeness' of Palestine including both banks of the River Jordan. The creation of the Kingdom of Transjordan was – and remains – in their view, the first partition of Palestine.

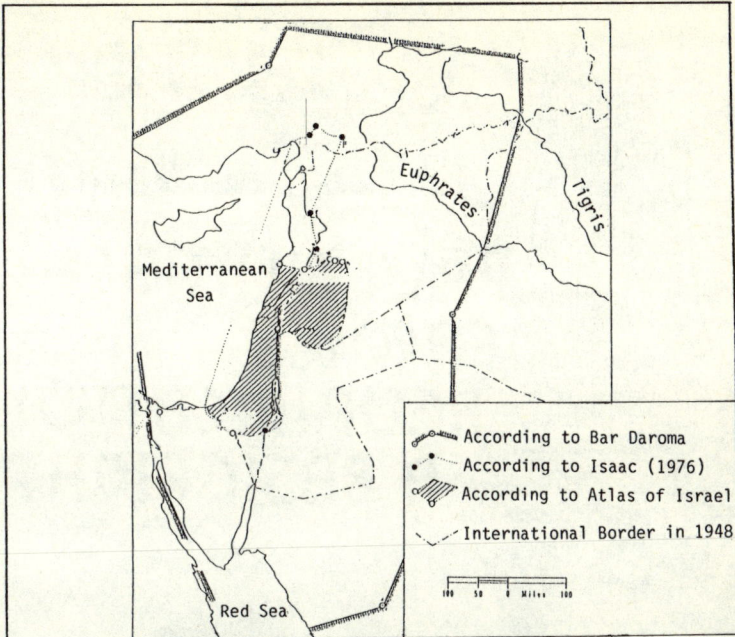

The different interpretations of the Biblical boundaries (taken from R. J. Isaac & E. Isaac, Israel Divided, 1976, Johns Hopkins UP).

The Israelite Kingdom under King David.

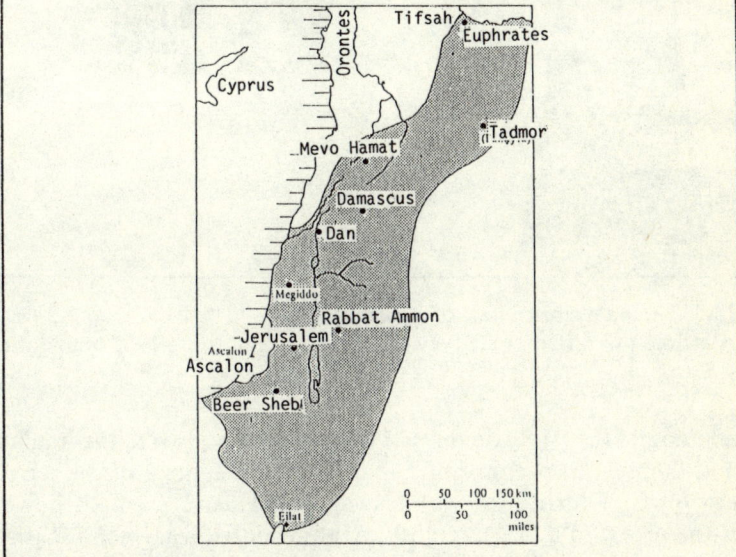

Figure 14.2 Maps showing the boundaries of 'Eretz Israel' according to biblical and historical periods.

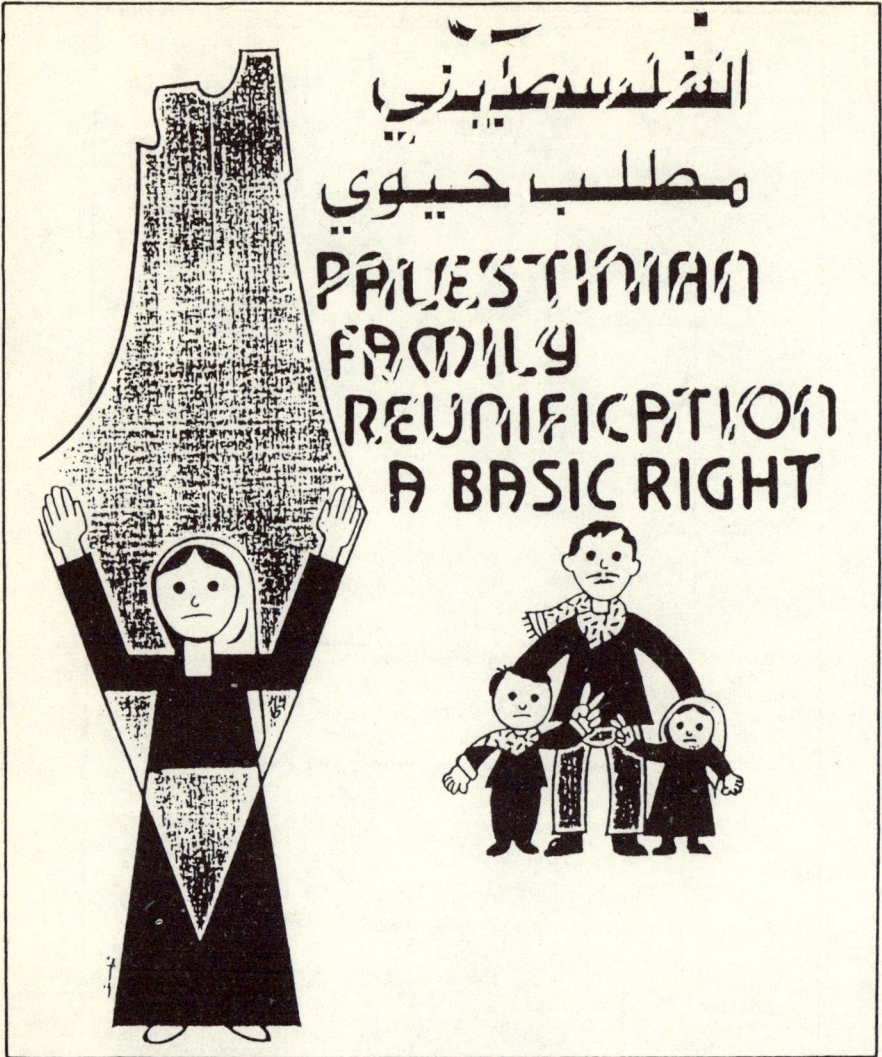

Figure 14.3 The Palestinian view of Palestine today.
   The map does not differentiate between Israel within its pre-1967 boundaries and the West Bank

powerful images, the strategic threat to Israel is emphasized, the implication being that there is no option other than to retain control of the territories since any form of territorial withdrawal would leave a small, powerless, Israel at the mercy of a large, mighty, Arab Middle East. Such images have been effectively used by right-wing political parties in their election campaigning. Their message – accompanied by appropriate maps – is that any territorial withdrawal or other concessions will result in national

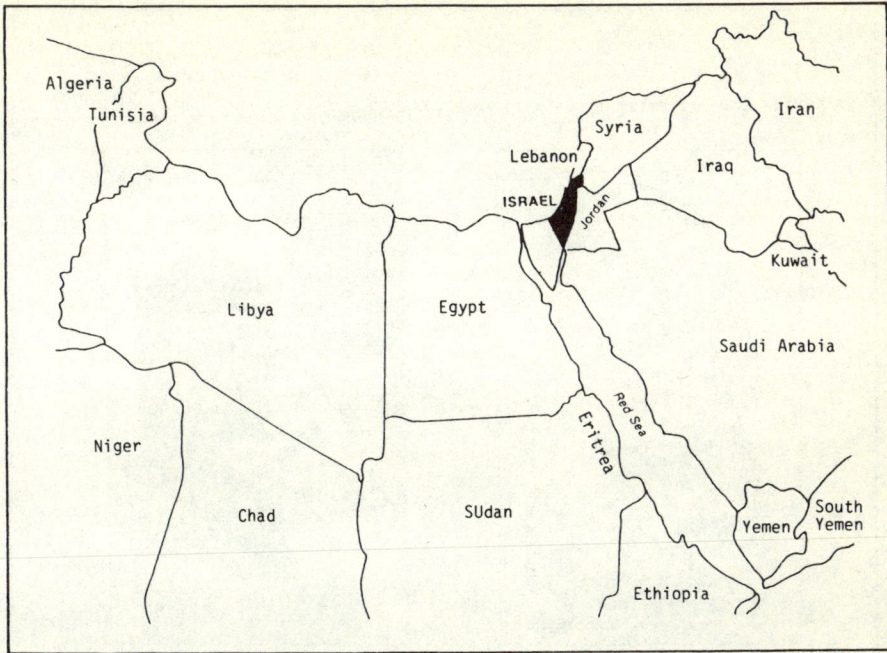

Figure 14.4 Real or imagined? The spatial dimensions of the maps showing Israel in the Middle East are true. However it is depicted in such a way as to emphasise a small Israel engulfed by an Arab Middle East.

suicide. A strong Israel is one that retains its current territorial advantages. Obviously, this image plays on the fact that some very real threats face Israel and that it is not always possible to distinguish between the extent of that threat and its exaggerated manipulation.

Place-name semantics and media images

The respective territorial images are also influenced by the terminologies used to describe the land in question. Hall (1981) notes that:

> the use/misuse of place/area names for purposes which are not always wholly altruistic, would appear to be particularly prevalent in irridentist situations where territorial claims and counter-claims, cultural diversity, and historical inter-digitations have lent particularly strong symbolic value to the use of certain spatial nomenclature (Hall, 1981: 320).

Such terminologies may be used both consciously and subconsciously, the latter arising out of long-term political socialization which is conditioned by

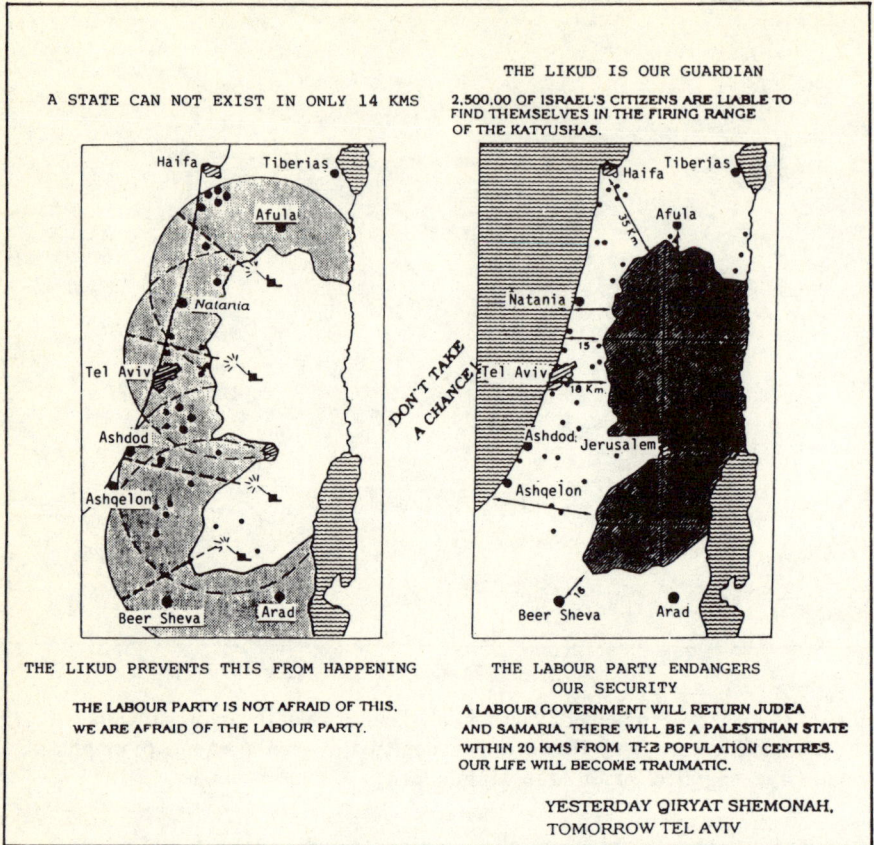

THE LIKUD IS OUR GUARDIAN

A STATE CAN NOT EXIST IN ONLY 14 KMS

2,500.00 OF ISRAEL'S CITIZENS ARE LIABLE TO FIND THEMSELVES IN THE FIRING RANGE OF THE KATYUSHAS.

THE LIKUD PREVENTS THIS FROM HAPPENING

THE LABOUR PARTY IS NOT AFRAID OF THIS. WE ARE AFRAID OF THE LABOUR PARTY.

THE LABOUR PARTY ENDANGERS OUR SECURITY

A LABOUR GOVERNMENT WILL RETURN JUDEA AND SAMARIA. THERE WILL BE A PALESTINIAN STATE WITHIN 20 KMS FROM THE POPULATION CENTRES. OUR LIFE WILL BECOME TRAUMATIC.

YESTERDAY QIRYAT SHEMONAH, TOMORROW TEL AVIV

Figure 14.5 Maps used by the Likud Party in the 1984 Election Campaign, showing the strategic threat involved in returning the West Bank to foreign rule.

These same maps were used by the Ministry of Foreign Affairs in their information campaigns around the world. Not surprisingly the Ministry was then in the hands of Likud leader, Yitzhak Shamir.

a process of territorial indoctrination (Shilhav, 1985). Specific names for the region, each with its own nationalistic and historical connotations, are used as a means of heightening the political consciousness of the non-committed. An analysis of these terminologies is largely structuralist in that it is concerned with the analysis of the mechanisms and processes whereby particular forms – in this case language and territorial terminologies – produce meaning.

In the case of the Arab–Israel conflict in general, and the West Bank in particular, the terminologies used by the media to define the territory under study further substantiate the socialization process of each of the national groups in the development of their environmental and territorial images

(Newman, 1988). No fewer than seven terminologies used to define the territory in question can be identified in the media and other literary sources (Newman and Portugali, 1987). These may be divided into two groups, the first of which gives a name to the region, the second adopting a politically biased functional usage.

The two main names given to the region are the West Bank and Judaea and Samaria. The former is the most generally acceptable terminology used by neutral onlookers and to a great extent by the Palestinian participants. Yet the term 'West Bank' is a relative one which defines a political unit established in 1948 as a result of the ceasefire between Israel and Transjordan. The 'Green Line' border became the unofficial state boundary between Israel and Jordan, i.e. the territory to the west of the River Jordan not under Israeli sovereignty. Within Israel, the term 'West Bank' was generally adopted before 1967, although there has been a gradual process of attrition in favour of Judea and Samaria during the last 20 years. Today, it is associated with centre and left-of-centre viewpoints, particularly among less chauvinistic territorialists.

The term 'Judaea and Samaria' is a more ancient label, used only by sections of the Israeli Jewish population. The terms 'Judaea' and 'Samaria' are the names of the two ancient independent Jewish kingdoms which existed in this region following the division of the Israelite monarchy after the death of King Solomon. The southern kingdom, with Jerusalem as its capital, was known as Judaea. The northern kingdom was initially known as Israel, but was later renamed Samaria after its newly built capital city, Samaria. Following the Six Day War, the territorial maximalists increasingly used this terminology in order to emphasize the connection between the region and the historical Jewish presence. This process of territorial indoctrination became legitimized following the formation of the first Likud-led government in 1977, after which the official State terminology for the region was changed to Judaea and Samaria. This terminology, increasingly used in Israel, is associated with the hawks camp, those wishing to retain control and eventually annex the region to the State of Israel. Under the government of Labour leader Shimon Peres, in 1985, his political advisor was reported as suggesting the insertion of the term 'West Bank' in brackets following the term 'Judaea and Samaria', in all official documents.

The second group of names in the literature comprise political-functional terms. Three major usages of this nature may be identified, ranging from 'occupied territories' through 'administered territories' to 'liberated territories'. The two extreme usages, 'occupied' and 'liberated', are perhaps far more indicative of political stances taken by the user than are the alternative use of West Bank or Judea and Samaria. The term 'occupied territory' is used by those who view the region as being under illegal occupation by an external force which has no right to remain there. Occupation is negative, it denotes an imposition of an unjust military rule which must be removed as early as possible. This term is also the generally accepted international functional usage. In the eyes of most outside powers, the West Bank is under occupation.

The alternative functional usage, 'liberated territories', is found less frequently. It is used almost exclusively by those groups who view the Six Day War as having brought about the rightful liberation of those territories belonging to the Jewish people and which had been, until then, under foreign (Jordanian) occupation or control. This term is used by far fewer Israelis than the term 'Judea and Samaria', indicating different intensities of belief that the land belongs to them.

The third, supposedly neutral, functional terminology is the 'administered territories'. This is generally used by government agencies (see for instance the quarterly report of the Israel Bureau of Statistics for the Administered Territories) and by those politicians, journalists, authors and scientists who do not wish to be overtly identified with any particular stance. A variation on this term is simply the usage 'territories' which, within Israel, has only one possible significance, i.e. those territories captured, conquered or liberated by Israel in the Six Day War of 1967. Another variation is 'beyond the Green Line', used only occasionally, and which denotes the region beyond the pre-1967 Green Line border.

A similar process of territorial differentiation takes place in regard to the names of specific places. Thus, Palestinians use Arabic names for villages and other settlements, while Israelis will use the Hebrew term. In many cases ancient Hebrew usage has been revived in order to stress the ancient Jewish links to the region, even in cases where only an Arab settlement exists today. Within pre-1967 boundaries, the Israeli government has changed the name of many existing urban centres in favour of a hebraized version. Beyond the green line, the town of Nablus (local usage) is referred to by its ancient Biblical name, Shechem, by Israelis in general, and settlers in particular. Similarly, the Arab town of Al Khalil (Arabic) is known by its biblical name of Hebron, while the important centre of Jerusalem (the international term is the Hebrew usage) is referred to as Al Kuds by the Arab world.

There is a conscious process on the part of the settlers to name the majority of the new settlements after the ancient Biblical or Israelite names for the region, thus promoting a consciousness of the Biblical-Hebraic attachments to the land. This changing of place names has taken place throughout Israel as a means by which to:

> re-establish contact with landscapes and places from which we had been physically removed for two thousand years but whose names we had always preserved . . . our geographica sacra (Benvenisti, 1987: 194).

More recent examples include the naming of the Jewish suburb of Hebron as Qiryat Arba, after its Biblical usage; the settlements of Beit El, Shiloh and Eilon Moreh (amongst others) after ancient Biblical and/or

Israelite usages. The collective use of the ancient Hebrew names has resulted from a long process of 'territorial indoctrination' by which the Jewish nation in exile retained its links with the far removed territory (Shilhav, 1985).

A similar parallel process of 'Palestinianization' also takes place. Palestinian refugees also have their *geographica sacra:*

> Their map making is the answer to our Hebrew Israeli map is as far removed from reality (Benvenisti, 1987: 198–9).

Pre-1948 Palestine is the base map for teaching the geography of the region, refugee camps are organized by sections according to villages of origin in Palestine, while the past geographical realities are often exaggerated and take the place of the truth in the formation of territorial images (Benvenisti, 1987). Territorial indoctrination is as strong a process here as it was for 2,000 years of Jewish diaspora.

Using images for peace?

Can images be used equally in drawing two conflicting participants together? Even attitudes towards the concept of peace are not uniform. Within Israel, there are many who do not view the peace with Egypt as constituting a 'real' peace. Some groups (especially among religious circles) view peace in a metaphysical sense of the lamb lying down with the lion and of swords being transformed into ploughshares. For others, such as the Peace Now Movement, peace implies neither more nor less than the cessation of hostilities and the commencement of bilateral negotiations (Newman and Hermann, forthcoming).

Spielmann (1986) has studied the attitudes to peace held by Arab and Jewish children from three different age groups: 9–10; 13–14; and 17–18. The survey was carried out twice, before and after President Sadat's historic visit to Jerusalem in 1977. In the earlier survey, Spielmann points to the generally more utopian and positive views of peace among the younger age groups, becoming more pessimistic with age, as images and perceptions become more deeply embedded in harsh realities. In the second survey, she found paradoxically that younger children were somewhat negatively affected owing to the discovery that peace meant 'giving up' something tangible, while the older children were on the whole positively affected.

> The Jewish children, primarily, wanted to guard what exists – the State. They saw peace as a secure situation in which anxiety and violence will be abolished, life will be pleasant . . . The Arab children saw peace as a complete change in their lives and the future as an obliteration of the past. Mostly, they envisaged the return of the Arab lands to the Arab people. Some looked for the establishment of a

Palestinian state besides Israel . . . But they did not perceive peace as a transformation in which everything was possible . . . (p. 64).

There has been little conscious use of terminologies or maps which are inherently constructive. Images depicting the possible gains to be made of cross-national development projects (for example; across the Arava border between Israel and Jordan (Gradus and Leibowitz, 1988; Minghi, 1988); a joint port and tourist centre between Israel and Jordan in Eilat; joint water management between Israel, Jordan, Syria and Lebanon; desert development schemes between Israel and Egypt) – while seeming far-fetched at this stage – could be effective.

While tangible attempts may be made to promote peace through negotiations, the ultimate 'normalization' of relations depends on long-term attitudinal changes, rather than on just a formal peace agreement and military disengagement. The continued use of negative images by each side for the other (such as the case of Israelis and Egyptians) results in continued distrust and a situation of 'cold peace'. While this is a vast improvement over a situation of 'war', the chances of renewed conflict will always be greater as long as attitudes remain unchanged.

If manipulation of media and other means of mass communication have become a norm in modern society, then it would be appropriate to use these means for promoting images depicting the opportunities to be gained from peaceful relations. Investment in programmes which increase each group's knowledge and awareness of the 'other' national group – their history, language and aspirations – would not be amiss. While a state of war can be transformed into a situation of ceasefire and peaceful negotiations within only a short period of time, changing attitudes takes generations. Yet in the long-term it is attitudes that are important if there is to be real normalization of relations between any two or more participants in a conflict.

Conclusions

I have done no more here than emphasize the importance of understanding image formation as a major obstacle in the transition from conflict to peace process. I have not offered a comprehensive survey of image formation. Investigation into the impact of both popular and scientific literature is another major means of communication which requires analysis. While popular literature will contribute to image formation through the characterization of people, groups and places in positive or negative terms, the impact of scientific literature is less clear. Yet Newman and Portugali (1987) have shown the difficulty in separating object from subject in supposedly 'objective' scientific analyses of the Israel–Palestinian conflict, noting the differential use of terminologies by the various writers – Israelis, Palestinians and outsiders.

Clearly, scientists have a role to play in promoting the use of alternative images – or at least not using negative images. Geographers should think carefully before attaching labels and terminologies to their maps and texts and should become more directly involved in the presentation of spatial images which display possible solutions to territorial and national conflicts. One application of geography to peace studies can be understood as a specific form of applied geography. In this way, political geographers can make a meaningful contribution in assisting politicians and decision-makers to arrive at conflict resolution.

References

Ager, J. (1978) Maps and propaganda, *Society of University Cartographers Bulletin*, **11**, 1–15.

Arian, A. T. I. and Hermann, T. (1988) *National Security and Public Opinion in Israel*. Tel Aviv University, Jaffee Centre for Strategic Studies, Study No. 9.

Benvenisti, M. (1987) *Conflicts and Contradictions*. New York: Villard Books.

Burgess, J. & Gold, J. R. (eds) (1985) *Geography, the Media and Popular Culture*. London: Croom Helm.

Burnett, A. (1985) Propaganda cartography. In D. Pepper and A. Jenkins (eds), *The Geography of Peace and War*. Oxford: Basil Blackwell, pp. 60–89.

Davies, Hunter (1984) *A Walk Along the Wall*. London: Weidenfeld & Nicolson (4th edition).

Downs, R. M. and Stea, D. (1977) *Maps in Minds*. New York: Harper and Row.

Gould, P. and White, R. (1974) *Mental Maps*. Penguin Books: London.

Gradus, Y. and Leibowitz, S. (1988) *The Israel–Jordan Rift Valley Initiative: A Proposal*. Beer Sheva:Ben Gurion University of the Negev, 4pp.

Hall, D. (1981) A geographical approach to propaganda. In A. D. Burnett and P. J. Taylor (eds), *Political Studies from Spatial Perspectives*. New York: Wiley, pp. 313–30.

Hudson, R. and Pocock, D. (1978) *Images of the Urban Environment*. London: Macmillan.

Minghi, J. (1988) From conflict to harmony in border landscapes. Paper presented to *The International Geographical Congress – Border Landscapes*. Perth, Australia, 15–19 August.

Moore, G. T. and Golledge, R. G. (1976) Environmental knowing: concepts and theories. In G. T. Moore and R. G. Golledge (eds), *Environmental Knowing*. Stroudsberg, PA: Dowden, Hutchinson and Ross: Stroudsberg.

Muir, R. and Paddison, R. (1981) *Politics, Geography and Behaviour*. London: Methuen.

Newman, D. (1988) *Mind over matter: images of the Arab-Israel conflict*. Survey of Arab Affairs, No. 14. Jerusalem Centre for Public Affairs, 5–8.

Newman, D. and Hermann, T. (forthcoming) Extra-parliamentarism in Israel: a comparative study of Gush Emunim and Peace Now.

Newman, D. and Portugali, J. (1987) Israeli Palestinian relations as reflected in the scientific literature. *Progress in Human Geography*, **11** (3), 315–32.

Pocock, D. and Hudson, R. (1979) *Images of the Urban Environment*. London: Macmillan.

Portugali, J. and Newman, D. (1987) Spatial interaction between Israelis and Palestinians in the West Bank and Gaza Strip. Unpublished Research Report, Ford Foundation Israel Trustees.

Relph, E. (1976) *Place and Placelessness*. Pion: London.

Romann, M. (1984) Arab and Jewish students images of Jerusalem: divided consciousness in a united city. *Horizons: Studies in Geography*, **17/18**, 77–104. (Hebrew).

Shilhav, Y. (1985) Interpretation and misinterpretation of Jewish territorialism. In D. Newman (ed.), *The Impact of Gush Emunim*. Croom Helm: London, pp. 111–24.

Spielmann, M. (1986) If peace comes . . . future expectations of Israeli children and youth, *Journal of Peace Research*, **23** (1), 51–67.

Waterman, S. (1979) Ideology and events in Israeli human landscapes. *Geography*, **64** (3), 171–81.

15 Geographical Perspectives on Peace Alternatives for the Land of Israel

Baruch A. Kipnis

Introduction

Geographical aspects of peace are usually perceived in terms of borders, security measures and landscapes, and in terms of the long-range impacts of the emerging bi-national normalization processes. However, very little research has been aimed at exploring the role of peace in shaping the spatial identity and the well-being of a community, be it an ethnic or a national group, remaining behind the peace borders in residual minority enclaves of a pluralistic multi-community state. This chapter, employing concepts derived from the literature on pluralistic states (for example: Esman, 1977; Hall, 1979; Lijphart, 1977; van Den Berghe, 1981), explores the above situation in the context of the Land of Israel, also referred to as Palestine, extending over 28,000 square kilometres between the Jordan River and the Mediterranean Sea. To this end, three issues are discussed:

1 Which are the most commonly declared peace arrangements aimed at accommodating Israeli Jews and Palestinian Arabs in the Land of Israel?
2 What would be the settlement landscape which would constitute a point of departure, or rather a deterministic factor, for any peace effort?
3 Using a 'policy goal split model', simulating inter-community spatial and functional interactions in a context of a pluralistic state or of a multi-community region (Kipnis, 1987b), the geographical outcomes of each one of the proposed peace alternatives is revealed and evaluated.

Declared peace alternatives

The kernel of the Jewish–Arab conflict over the Land of Israel, thus possibly the only basis for its peaceful solution, is a territorial one. The national territory is a multi-dimensional entity in which a national community is secure and is allowed to operate its institutions (Burghardt, 1969; Reichman, 1973; Kipnis, 1987a). Both Arabs and Jews believe that they possess the sole historical rights over the entire territory of the Land of

Israel. At best, each would offer the other subordinate group minority status.

Assuming the above prerequisites along with the mere fact that reality also ought to play a role in determining the fate of the Land of Israel, three scales of peace alternatives are proposed: Maximum (MAX); Medium (MED); and Minimum (MIN). These scales, coupled with the two communities, create six declared 'peace' solutions, three by Arabs and three by Jews (Table 15.1).

The Arabs who insist on a MAX solution aim at the one proposed by the 'Palestinian Covenant'. This covenant calls for the establishment of a Secular Democratic state with an Arab majority and a small Jewish minority. This State is proposed for the entire land territory of Palestine (the Land of Israel). The Jews for their part, while demanding a MAX solution, claim that the entire Land of Israel should be declared an integral part of the Jewish State of Israel.

The MED scale Jewish alternative envisions a solution according to the Allon Plan of late 1960s. Allon, a notable statesman and a security expert, was Minister of Education when he proposed his plan. He later became the Minister of Foreign Affairs. The Allon plan proposed that the Jordan Rift Valley should become a defensive buffer zone between Israel and the Eastern Front of belligerent Arab States, including Iraq, Syria, Saudi Arabia and Jordan. Similarly, the region of Jerusalem, including Arab East Jerusalem and a few pockets on the edge of the Samarian and Judea hills (the West Bank), should remain in Jewish hands. The hard core of the Arab region in mountainous Judea and Samaria and in Gaza, should be returned to Jordan and to Egypt as part of a peace settlement. (In the peace treaty with Egypt of the late 1970s, the Egyptians refused to take back the Gaza Strip with 500,000 Palestinians under their jurisdiction.) The Allon Plan also called for the imposition of security measures on the Territories. These included a ban on the entry and the placement of military forces and equipment, and the imposition of strict demilitarization arrangements (Yad Tabenkin, 1984).

The Arabs' MED scale alternative, aired once more in recent months, suggests a revival of the 1947 UN Partition Plan, a plan that the Arabs originally had rejected. Under this MED Arab alternative, Israel would lose some 5,000 square kilometres of its pre-1967 borders' land territory. Notable among the regions that would be lost are the mountainous Galilee and Jerusalem. The former would become part of the Arab State while Jerusalem would become an international city. By accepting the Arabs' MED plan, two independent States would be created in the Land of Israel: an Israeli Jewish and a Palestinian Arab.

The MIN scale alternatives, both of the Jews and of the Arabs, advise the return of Israel to its pre-1967 Six Day War boundaries. This is the only solution with support and agreement among moderate Jews and Arabs on its basic territorial dimensions and principles.

Table 15.1   Peace alternatives by initiator and scale

Scale of Alternatives	Initiators	
	Arabs	Jews
Maximum (MAX)	Peace arrangements under the terms of the 'Palestinian Covenant' – a 'secular democratic' state with Arab majority and Jewish minority	Jewish sovereignty over the entire territory of the Land of Israel. The Arabs would be granted an Autonomy in the West-Bank and in Gaza.
Medium (MED)	The return of Israel to the borders proposed by the 1947 UN 'Partition Plan', and the creation of two independent national States.	Arrangements according to the 'Allon Plan', proposing Jewish sovereignty over the Jordan Rift Valley, over the region of Jerusalem, and over a few pockets in and around the West-Bank. Areas with large Arab population in the West Bank and Gaza should be returned to Jordan. Tough security measures should be imposed on the above regions.
Minimum (MIN)	The return of Israel to its 1967 pre-Six-Day-War borders and the creation of two independent national States	The return of Israel to its 1967 pre-Six-Day-War borders except for the region of Jerusalem, the Golan Heights, and a few small border segments at the peripheries of the West-Bank. The remaining territories should either be returned to Jordan or be designated as the land territory of a Palestianian State.

The settlement landscape: a framework for peace arrangements

Due to the geopolitical process of evolution of the settlement space during the last 100 years, and more so since the mid-1970s, none of the above peace solutions for the Land for Israel would result in a complete territorial separation between Jews and Arabs. Throughout the entire period of the geopolitical process of settlement formation in the land of Israel, the Jews, oriented by their geopolitical ideologies, have been active in crystallizing and implementing their territorial aspiration by means of 'Macro' scale development strategies, nested in the eternal notion that the Land of Israel is the sole national 'homeland' of the Jewish people.

Jewish territory-forming ideologies have been as follows (Kipnis, 1987 a, b):

1 A spatial deployment of rural settlements, an ideology inspired by the belief that settlements ensure control over the national space and over its viable resources, and that settlements constitute an essential element in national security.
2 A population balance ideology, attained by urbanization policies, assuming that Jews should constitute a majority throughout the entire national territorial space. The Galilee New Towns of Upper Nazareth, Ma'alot and Karmiel, and the Samarian New Town of Ariel, are the leading examples of the employment of the population balance strategy.
3 The 'deterritorialization' of the Arab settlement space ideology, implemented by means of territorial isolation and institutional dependency strategies, generated from the notion that within the territory controlled by the Jewish community, the extent of which has changed over time, there is room for only one national territory, a Jewish one.

The Arabs on the other hand, have revealed mainly passive attitudes or attempted to impede development through political action or violence. From 1948 to the mid-1960s, being under Military Administration, the Arabs were confined to passive spatial and political roles only; since the mid-1960s, however, when the Military Administration was abolished, and more significantly since the mid-1970s when the induced Jewish settlement deployment moved into the mountainous regions of Galilee, Judea and Samaria, and into Gaza strip, the Arabs, concerned at their diminishing space, have striven to employ small-scale spatial strategies.

These strategies, aiming at 'protecting' their territory to its utmost possible limits, included 'latent' urbanization (Meir-Brodnitz, 1969; Kipnis, 1976); illegal housing construction (the Kovarski Commission Report, 1976; The Markovitz Commission Report, 1986); extensive 'sedentarization' of Bedouins (Medzini, 1986); superficial land reclamation and the planting of olive trees at the peripheries of their villages. The end result of both the Jewish large-scale and of the Arabs small-scale strategies has been a 'mixed settlement space', to be found in almost every region of the Land of Israel (Soffer, 1986). Included here are mixed Jewish–Arab cities (Kipnis and Schnell, 1978; Ben-Artzi, 1980; Ben-Artzi and Shoshani, 1986). This space is referred to as 'mixed community' (Reichman, 1980), as a 'duality space'

(Grossman, 1987); and as 'implicative' or 'contained' space (Portugali, 1986). All result in a growing conflict situation which might, if not eased, increase inter-community tension. This tension has already inspired strong irredentism and national claims for self-determination by the Palestinian Arabs. The Arab uprising ('intifada'), and the recent autonomy claims by the Israeli Arabs, are but a few signs in this direction.

Evaluation of peace alternatives – the IE model

Among the many models proposed for a multi-community state (Rabushka and Shepsle, 1972; van den Berghe, 1981; Smooha, 1982), three seem to have wide support: a 'Liberal Democracy', a 'Consociational Democracy' and a 'Domination Democracy'. Their attributes reflect the legitimacy assigned to inter-community relationships, whose scope is confined by the degree in which one community monopolizes powers and privileges, and by the extent to which citizens are incorporated into the state (van den Berghe, 1981). The legitimacy assigned to the spatial organization policies in a pluralistic state are bounded by the state model attributes. The scope of inter-community relationships under the first legitimacy (the degree of manipulation of power) extends from a situation in which conflicting communities strive to maintain a negotiated co-existence as in the Consociational Democracies of Belgium and Switzerland, to states in which a minority community fully dominates the majority, as in South Africa and in pre-independence Zimbabwe. In between there are several pluralistic states like Israel, in which one community possesses domination status, but it is quite open to share resources, and to grant free and equal citizenship rights and community cultural autonomy to its minority community. The scope of the second legitimacy varies between states in which citizens are fully incorporated according to their individual merit, as in the Liberal Democracy of France, to states in which communities are legally recognized and granted different rights and obligations. Development policies and strategies assume different courses of action when they are applied in pluralistic states according to their model affiliation.

The leading policy of a 'Liberal Democracy' is 'consensus building', aimed at providing equal and just accessibility to resources and services, on the basis of personal merits (Smooha, 1982). In so doing, the policies strive to eliminate ethnic division, to weaken primordial ties, and to induce shared values and common identities. In spite of the fact that a 'Liberal Democracy' officially recognizes only individuals, it might tolerate some degree of either imposed or voluntary pluralism (van den Berghe, 1981). When this occurs, inter-community conflicts are resolved, or rather confined, to local enclaves or to small regions. A prominent development policy for a 'Consociational Democracy' is to induce inter-community solidarity by means of a 'negotiated compromised co-existence'. Under the compromised co-existence policy each community operates within an agreed cultural and territorial autonomy; political integration and proportional inter-community representation are maintained in most national institutions. Major efforts are made to crystallize a broad operational consensus, to compromise on controversial issues and to depolarize inter-community disputes. Development policies are further employed to enhance fair allocation of resources and to promote equal opportunities and social mobility (van den Berghe, 1981; Smooha, 1982).

'Consociational Democracies' are among the most vulnerable of states. Severe regional and inter-community inequalities have led, in some cases, to a continuous struggle over limited resources, resulting in irrational and inefficient allocation and management of limited national wealth. If this occurs, latent or open aspiration for secession may prevail, leading to growing claims for federalism, as in the case of Belgium, or to territorial fragmentation, as in the case of Lebanon (Rabushka and Shepsle, 1972). With the objective of reducing the likelihood of secession and in order to preserve a stable consociational framework, some policy measures should be implemented. These are that the constituent communities should not be hierarchically organized in terms of wealth and prestige; that cultural pluralism remains relatively stable; that in spite of the inherent territorial separation, some degree of spatial, genetic and functional interdependencies and mix must prevail; and that strong central government and a vivid mixed capital be given the utmost attention.

The leading policy measure for maintaining social stability in a 'Domination Democracy' is that the dominating community possesses control over the national space and over its resources, infrastructure and institutions (Smooha, 1982). If geopolitical considerations are also involved, as in the case of the Land of Israel, the dominating community tends to deterritorialize, by means of regional strategies, the territory of the other community. Regardless of the policy measures employed by the dominating community, there are increasing indications that the 'Domination Democracies' are moving slowly towards a 'Consociational' or a 'Liberal' Democracy' frameworks (van Den Berghe, 1981). However, the rate of change and its direction depend on the real impact of the imposed development measures, on the dynamics of social, economic and political transformations that take place in the subordinate community, and on the changing degree of tolerance by the dominating one.

Of the three types of pluralistic states, the Jewish–Arab relationships in the Land of Israel best resemble the characteristics of a 'Domination Democracy'. In this context the Jewish community has been dominant since the creation of Israel as an independent State in 1948. In order to reveal and evaluate the geographical perspectives of the peace alternative for Israel, an Integration-Equality (IE) 'policy goal split model' is proposed (Figure 15.1). The model delineates among alternatives across two interrelated policy goal

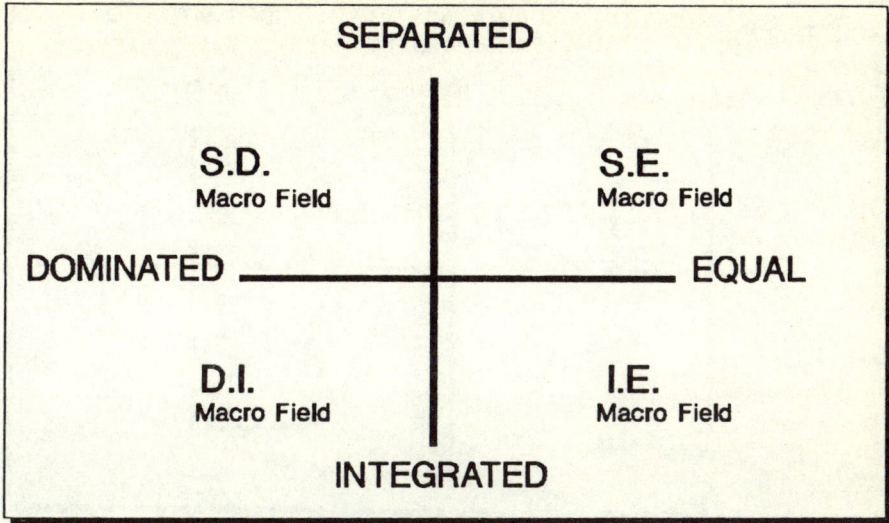

Figure 15.1 A policy goal split model for a pluralistic state/region and its Macro-Fields

stretched continua: 'Equal'- 'Dominated' and 'Integrated'-'Separated'. These stretches divide among four Macro-Fields:

1 A 'Separated-Equal' (SE) Macro-Field, in which two or more independent and equal national communities co-exist;
2 A 'Separated-Dominated' (SD) Macro-Field, where two or more unequal communities prevail, one of which dominates the resources and the state of affairs of the other(s);
3 An 'Integrated-Dominated' (ID), simulating a situation in which two or more national communities are 'forced' to integrate in space and in functions by means of regional development strategies. In an ID State situation, the dominating community determines which are the functions and the regions in which inter-community integration should take place, as well as its timing and rate of its integration processes. In addition, the dominating community possesses the powers enabling it to control and to ban access to resources and to opportunities from the subordinate one; and
4 An 'Integrated-Equal' (IE) Macro-Field, allowing all national communities to enjoy equal and free access to resources, opportunities and to all shared national and regional institutions.

Each of the above Macro-Fields, it is argued, might suit a different type of pluralistic State. The SE Macro-Field, implying a well-defined spatial cultural separation between the communities involved, best suits the

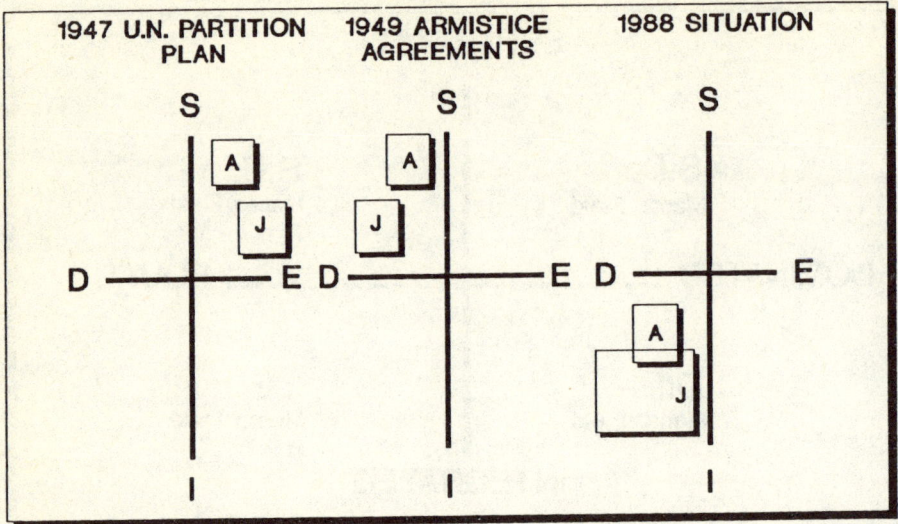

Figure 15.2 Evolution of Macro-fields' settlement patterns in the Land of Israel

'Consociational Democracy' State; the IE Macro-Field on the other hand, possesses the essentials of a 'Liberal Democracy'; the SD and the ID Macro-Fields simulate the various situations of spatial and functional affairs of a 'Domination Democracy'. As indicated earlier, the SD and the ID Macro-Fields represent the existing inter-community relationships in the mixed territorial enclaves in the Land of Israel (Figure 15.2). Note how the outcomes of the 1948 War of Independence resulted in an SD Macro-Field settlement pattern. It included the Jewish territory evolved during the pre-statehood years along with the territory taken during the war. This land territory, recognized by the 1949 Armistice Agreements as Israel's, had some 850,000 Jewish and 156,000 Arab inhabitants. The Arab community of post-1948 war has been either residual of a larger community who left during the war, or those who lived in a territory ceded to Israel by Jordan as part of the Armistice Agreements. The hard core of the Arab territory was in the mountainous Galilee region and in the 'Triangle' region, extending along the 1949 border with Jordan. Since then, close to 75 per cent of the Israeli Arabs have lived in these two regions.

Implicit in the 'policy goal split model' is that each one of the Macro-Fields might yield one or several Micro-Field(s). An SE Macro-Field, for example, might contain one or a combination of Micro-Fields such as SE and ID. Using the IE model, both the Arab and Jewish MAX peace options would result in only one State being located for a very long period of time at the ID Macro-Field (Figure 15.3). The MED and the MIN peace alternatives, on the other hand, would end up assuming that tension has been eased and conflicts resolved, in two states situated at the SE Macro-

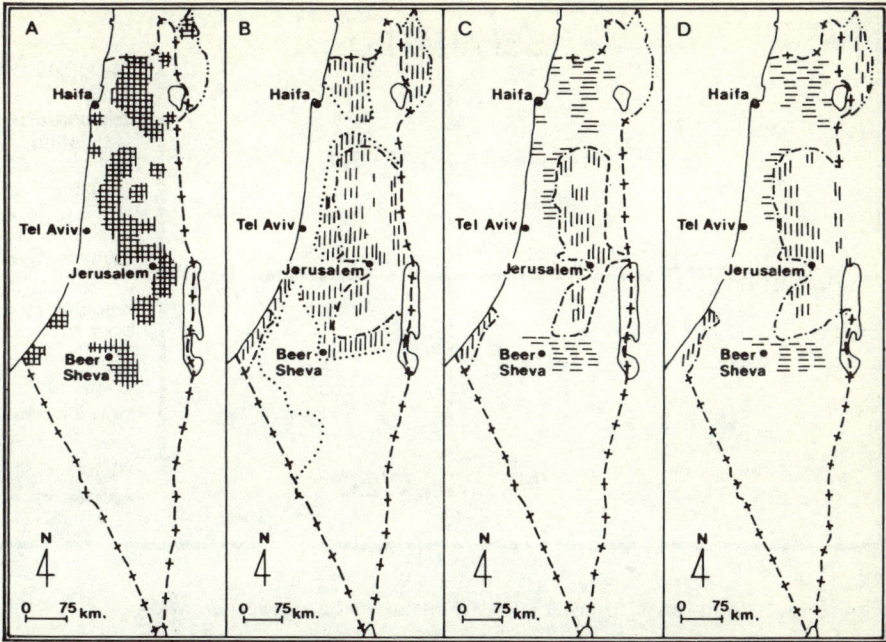

Figure 15.3 Alternative territorial peace arrangements for the Land of Israel

Field. Each one of the independent states, however, would possess, at least during its early stages, several regions possessing one or a few ID Micro-Fields (Figure 15.3).

At issue is whether the ID Macro and Micro-Fields, the outcome of the above MED and MIN peace alternatives, would change into an IE Macro and Micro-Fields over-time. Of similar importance is the question whether all of the functional systems of the Jewish and of the Arab communities would follow the same patterns of change? As indicated earlier, van den Berghe (1981) suggests that there are increasing indications that the 'Domination' democracies are moving slowly towards a state of 'Con-sociational' or 'Liberal' democracies. The rate of change and its direction, it is argued, depend on the dynamics of the social, economic, cultural and political change that would take place within the communities involved. Due to the mixed enclave situation that has evolved in the Land of Israel over time, the direction of change towards a 'Consociational Democracy' State, which assumes spatial, cultural and institutional separation, ought to be ruled out.

The only possible course of change of the emerging multi-community states of the Land of Israel is towards the direction of a 'Liberal Democracy' entity. In this framework the communities involved would strive to increase their consensus over major issues, and would attempt to enjoy improved

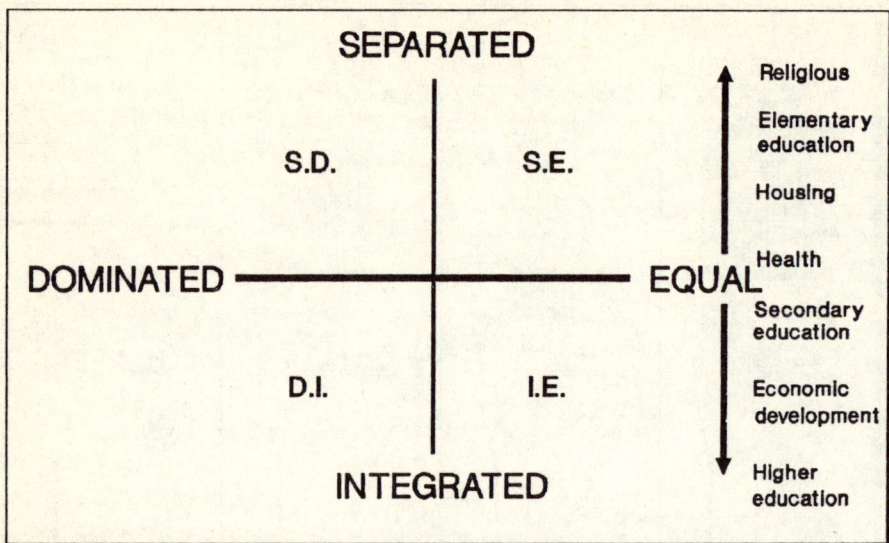

Figure 15.4 Long-range Micro-Field arrangements of a peace situation

access to resources, opportunities and to institutions. Drawing once more from van den Berghe (1981), and implying that, in spite of the fact that 'Liberal Democracies' try to work hard to eliminate ethnic division, to weaken primordial ties, and to induce shared values and common national identities, it also might tolerate some degree of either imposed or voluntary pluralism. This seems to be the situation in the case of Arab–Jewish relationships in the Land of Israel in which the two communities would maintain, in the long run, many of their national and cultural identities. Although it is difficult to predict the actual courses of change, a few scenarios as to the spatial organization of a peaceful settlement space are drawn. Stretching along the Equal continuum of the IE model (Figure 15.4), some functions would probably remain at their S (separated) edge. These might be religious and cultural services, and possibly elementary education. Housing and secondary education might evolve into some degree of a mix, and they would probably locate at the midway between the S and the I edges of the Equal stretch. Closer to the I end of the Equal stretch one may find some 'culture-free' functions such as health, economic development and higher education. In all of this latter group of functions, the level of day-to-day cooperation between Arabs and Jews has been relatively high during the last two decades. The only expected change to occur in a situation of peace over the Land of Israel, is that inter-community relationships would move from the ID Macro-Field into the IE one. In the latter, various degrees of culturally-bounded and culture-free activities would exist, while each of the communities involved would strive to maintain its own identity.

Concluding remarks

In this chapter an attempt has been made to speculate on the dimensions, nature and on the main attributes of Arab–Jewish relationships in a situation of peace in the Land of Israel. The outcome of the analysis is that regardless of the peace alternative to be implemented, Jews and Arabs would have to continue to live in a minority–majority relationship for many years to come. How do communities interact in a situation of 'minority–majority' relationships in a pluralistic state? To this end, the concepts of *cohesion*, namely sticking together, and *covariance*, i.e. changing together, coined by Deutsch (1974) for a system interaction analysis are employed. The degree of cohesion and of covariance reflects the system's interdependencies, manifested by its dispersal of goods and services and by the movement of its people. The flow of transactions, the outcome of a given state of inter-community relationships, might yield either positive or negative covariance of rewards. Reward is positive if the system of relations and of interactions is aimed at increasing the values enjoyed by all communities. The rewards are negative if the system of interactions, either by intent or as a result of malpractice or of bad policies, reduce the utilities received by one of the communities. Negative covariance of rewards induce dissatisfaction and conflict. Positive rewards, on the other hand, promote solidarity, co-existence and mutual responsibility. Needless to say, the former rewards represent the existing situation; the latter is expected to evolve in a situation of peace.

It is very hard to predict how long it takes before real peace and positive rewards would arrive in this region. The conflict is extremely vulnerable, loaded with emotions and conceptions and, even more so, often with unrealistic aspirations. Our task as geographers is to try to draw a reasonable framework for a better understanding of the spatial outcomes in each of the peace alternatives.

References

Ben-Artzi, Y. (1980). *Residential Patterns and Interurban Migration of Arabs in Haifa*. Occasional Papers on the Middle East, New Series, Haifa: University of Haifa (Hebrew).

Ben-Artzi, Y. and M. Shoshani (1986). *The Arabs of Haifa 1972–1983: Demographic and Spatial Changes*. Haifa: The Jewish Arab Center University of Haifa (Hebrew).

Burghardt, A. (1969). The core concept in political geography: a definition of terms, *Canadian Geographer*, **13**, 349–59.

Deutsch, K. W. (1974). *Politics and Government, How People Decide their Fate*. Boston: Houghton Mifflin, 2nd edition.

Esman, H. J. (ed.) (1977). *Ethnic Conflict in the Western World*, Introduction. London: Cornell University Press.

Grossman, D. (1987). Unified and dual space: the process of Jewish settlement in Western Samaria, *Horizons in Geography*, **21**, 27–60 (Hebrew).

Hall, R. L. (1979). *Ethnic Autonomy-Conspirative Dynamics*. New-York: Pergamon.

Kipnis, B. A. (1976). Trends of the minority population in the Galilee and their planning implications, *City and Region*, **3**, 54–68. (Hebrew).

Kipnis, B. A. (1987a). Geopolitical ideologies and regional strategies in Israel. *Tijdschrift voor Econ. en Soc. Geografie*, **78** (2), 125–38.

Kipnis, B. A. (1987b). Regional development and strategy considerations in multi-community Land of Israel. In Hofman, J. H. (ed.), *Arab-Jewish Relations in Israel*. Bristol, Ind. Wyndham Hall Press, pp. 21–44.

Kipnis, B. A. and I. Schnell (1978). Changes in the distribution of Arabs in mixed Jewish-Arab cities in Israel, *Economic Geography*, **54**, 167–80.

Kovarski, H. (1976). Illegal housing construction in the minority sector. Submitted by the Director General of the Ministry of the Interior to a Ministerial Committee 28/6/1976 (memo) (Hebrew).

Lijphart, A. (1977). *Democracy in Plural Societies*. New Haven: Yale University Press.

Markovitz Commission Report (1987). *Illegal Construction in the Arab Sector*. Submitted to the Government of Israel. (Hebrew).

Medzini, A. (1986). *Distribution of Bedouins in the Galilee as a Result of a Spontaneous Sedentarization and of Government Policy*. MonoGeoGraphy Series in Geography. Haifa: Department of Geography, University of Haifa (Hebrew).

Meir-Brodnitz, M. (1969). Latent urbanization in Arab villages, *Environmental Planning*, **89**, 4–12 (Hebrew).

Portugali, Y. (1986). Labor fields of Arab laborers in Tel-Aviv: preliminary findings and workable hypotheses, *Horizons in Geography*, **17–18**, 25–48.

Rabushka, A. and A. Shepsle (1972). *Politics in Plural Societies*. Columbus: Charles Merrill.

Reichman, S. (1973). The political geography of the Israel–Arab conflict, *City and Region*, **1**, 41–6 (Hebrew).

Reichman, S. (1980). Geography and politics in Jersualem, *Abstracts of the Annual Meeting of the Israeli Association of Geographers*, Jerusalem, 51. (Hebrew).

Smooha, S. (1982). Existing and alternative policy towards the Arabs in Israel, *Ethnic and Racial Studies*, **5**, 71–98.

Soffer, A. (1986). The territorial conflict in Eretz-Israel, *Horizons in Geography*, **17–18**, 7–24 (Hebrew).

van den Berghe, P. L. (1981). *The Ethnic Phenomenon*. New York: Elsevier.

Yad Tabenkin (1984). *Allon Plan, Supplement to Maps*. Institute for Defence Forces Studies. Tel-Aviv: Yad Tabenkin (Hebrew).

Index